KU-436-478

GUIDE TO THE
Wildflowers
OF SOUTH WESTERN AUSTRALIA

PHOTOGRAPHY BY SIMON NEVILL
TEXT BY SIMON NEVILL & NATHAN McQUOID

SIMON NEVILL PUBLICATIONS

CONTENTS

INTRODUCTION

The south west region of Western Australia has one of the richest floras on Earth, a very ancient landscape supporting around 9000 plant taxa. This makes it a very productive and rewarding place to visit for those who have an interest in nature, especially wildflowers.

The book has been designed as an introductory guide to the wildflowers of the South Western Botanical Province. It attempts to give the reader several opportunities to understand and identify the many wonderful plants of the south west, as well as being a guide to where to find the 'flora rich' areas. With over 900 species illustrated there is ample reference material to assist the reader in identifying many of them.

The first chapter shows where and when to find the wildflowers in the south west. It contains maps to aid travel planning and highlights some of the 'flora rich' areas. It also mentions possible places to stop, look and find particular plants. As you travel through the south west you will pass through the different botanical zones and some species will be the same and others totally different. This is what makes the study of plants here so fascinating.

The second chapter discusses some of the reasons why there is such a wonderful diversity of plants and explains the use of botanical zones to describe the South West Botanical Province. It then gives a brief overview of plant structure and shows some of the major wildflower families. A brief section on growing native plants has been included as well as on pollination and honeybees. It also discusses the problems that affect our wildflowers and the importance of conservation as a means to preserve Western Australia's unique flora.

The third chapter is the largest section of the guide book, containing photographs of over 900 plant species, in fact more than any other book to date on wildfowers of the south west. It also contains detailed sketch maps of selected reserves and parks to visit. The chapter has been divided into 7 primary botanical zones that are confined to the South West Botanical Province and includes an additional zone that is outside this region, to assist the traveller in the search for various everlasting species. The south west zone includes: the Karri-Tingle Forest, the Jarrah-Marri Forest, the Banksia-Eucalypt Woodland, the Wandoo-Eucalypt Woodland, the Semi- arid Eucalypt Woodland, the Northern Mallee Shrublands and Heath and the South Mallee Shrublands and Heath, and the additional section, the Mulga Woodland.

We hope this book makes your journeys through the 'Wildflower State' enjoyable, be it by car or just sitting back in your armchair. Enjoy.

Chapter 1

Where and when to find Wildflowers in the South West

The South West Botanical Province covers only 15% of Western Australia's total land surface area but contains more species of plant life than the rest of the state, so your travels in this region should be very rewarding.

Some specific areas and plant comunities contain more species than others so if you were to say, visit the sandheaths called Kwongan from the aboriginal Nyoongar word for sandy plain, you would find a vast array of wildflowers in season, but if you were in the tall Karri forest of the deep south west you would find less wildflowers, although of course you would be in a most beautiful area well worth visiting.

To assist in finding those routes that have a greater diversity of wildflowers, the roads are highlighted in red, but we strongly recommend that you try and visit as many vegetation zones as possible to gain a greater understanding of why one area is richer than another.

"When and where should I travel to see the wildflowers in the South West?" is the big question most people ask. The rest of this chapter will describe where to find the most prolific wildflower areas, but the 'when' is not so easy because different regions can have different flowering times and can also be affected by rainfall. Generally, wildflowers are at their most prolific between the months of July and November.

When you become a real wildflower enthusiast and know what is flowering when and where, you would never restrict yourself to just one period of the year. Take for example the verticordias, called feather flowers. They are at their best in November and December north of Perth and even later in the south. Many of the Banksias and Eucalypts are flowering from January to March and some of our wonderful orchids are out from May to July, so there is always something flowering somewhere at some time.

If you are travelling from interstate or overseas then your time here may be precious and you obviously will want to see as many species as possible. Between August and October should be the most rewarding time, starting north of Perth first and slowly making your way to the south with the deep south west corner last. If the sight of carpets of everlastings is important for you , then this can be as early as the beginning of August and those areas are highlighted in the map 'C', but the intensity and timing is related directly to how good the preceding winter rains were. Sometimes lack of rain can affect this region and there will be virtually no everlastings then, but don't despair, there will always be flowers blooming somewhere in the south west. If your time is limited, we have shown areas close to Perth that can be seen in a day, but the longer you have to stay the greater the variety of wildflowers you will see. The maps extend to the limit of the South West Botanical Province.

You will read about important botanical areas to visit in this chapter as shown on the main regional maps. There are also sketch maps of particularly important reserves located in the appropriate botanical zone under Chapter 3 'The vegetation zones of the South West Botanical Province'.

The maps are a guide only and we recommend that you obtain more detailed maps of each area from map shops, motoring organisation or service stations and for details on the various parks and reserves. The State Government agency, the Department of Conservation and Land Management has very good maps on the various National Parks and Nature reserves. They should be able to assist with what information you may require.

For accommodation details the Western Australian Tourism Commission can assist. They can also help you with reputable tour companies that run wildflower tours. There are benefits with travelling with local guides as they know their area well and should be able to assist with those tricky species that you will not know, but don't expect them to know all of them as there are more than 8,000 in the South West Botanical Province.

Just a few reminders...

Some of the roads noted on the maps are dirt roads that require a different type of driving technique, so drive carefully. Also some of them are in remote areas, particularly due east of Hyden and Lake King, and vehicles should be in good working order and water and other emergency equipment should be carried. You should notify people of your plans and return dates. Also in the more populated areas be careful when leaving vehicles and make sure that your vehicle is locked and valuables either with you or out of sight; theft is not common in parks but has occurred from time to time. Last but not least, the picking of wildflowers in the whole of Western Australia is prohibited.

Well, enjoy your visit to the South West and happy wildflower travelling.

THE VEGETATION ZONES OF THE SOUTH WEST BOTANICAL PROVINCE

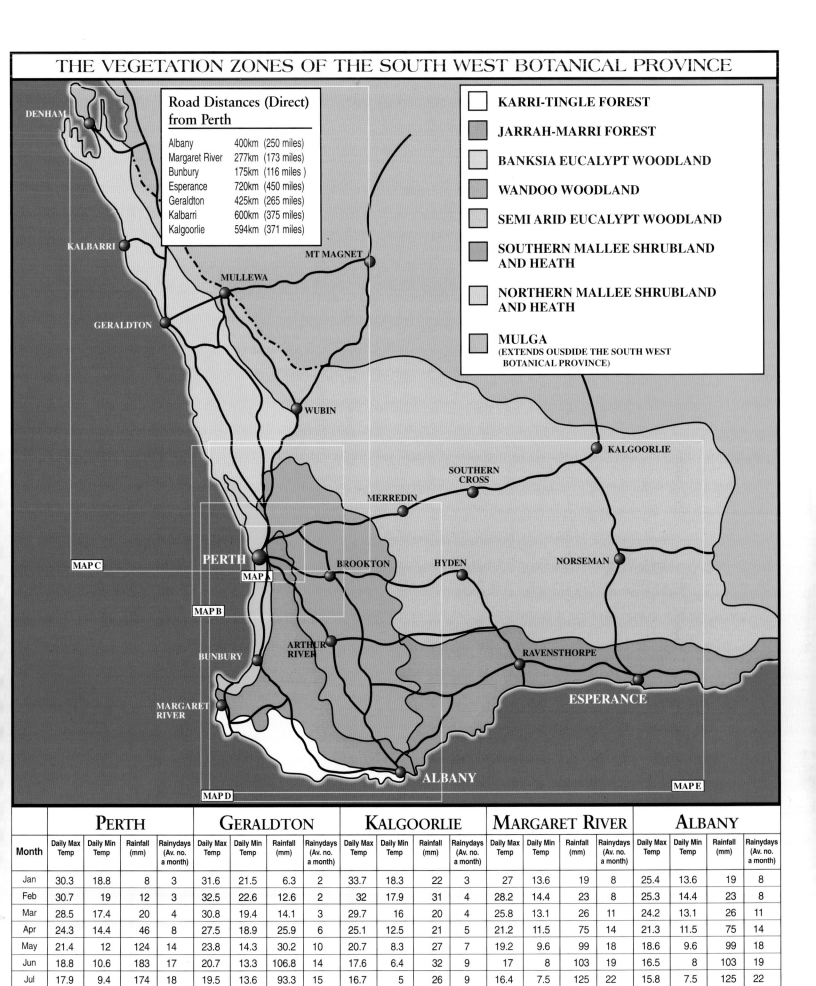

Road Distances (Direct) from Perth

Albany	400km	(250 miles)
Margaret River	277km	(173 miles)
Bunbury	175km	(116 miles)
Esperance	720km	(450 miles)
Geraldton	425km	(265 miles)
Kalbarri	600km	(375 miles)
Kalgoorlie	594km	(371 miles)

Legend:
- KARRI-TINGLE FOREST
- JARRAH-MARRI FOREST
- BANKSIA EUCALYPT WOODLAND
- WANDOO WOODLAND
- SEMI ARID EUCALYPT WOODLAND
- SOUTHERN MALLEE SHRUBLAND AND HEATH
- NORTHERN MALLEE SHRUBLAND AND HEATH
- MULGA (EXTENDS OUSDIDE THE SOUTH WEST BOTANICAL PROVINCE)

Month	PERTH Daily Max Temp	PERTH Daily Min Temp	PERTH Rainfall (mm)	PERTH Rainydays (Av. no. a month)	GERALDTON Daily Max Temp	GERALDTON Daily Min Temp	GERALDTON Rainfall (mm)	GERALDTON Rainydays (Av. no. a month)	KALGOORLIE Daily Max Temp	KALGOORLIE Daily Min Temp	KALGOORLIE Rainfall (mm)	KALGOORLIE Rainydays (Av. no. a month)	MARGARET RIVER Daily Max Temp	MARGARET RIVER Daily Min Temp	MARGARET RIVER Rainfall (mm)	MARGARET RIVER Rainydays (Av. no. a month)	ALBANY Daily Max Temp	ALBANY Daily Min Temp	ALBANY Rainfall (mm)	ALBANY Rainydays (Av. no. a month)
Jan	30.3	18.8	8	3	31.6	21.5	6.3	2	33.7	18.3	22	3	27	13.6	19	8	25.4	13.6	19	8
Feb	30.7	19	12	3	32.5	22.6	12.6	2	32	17.9	31	4	28.2	14.4	23	8	25.3	14.4	23	8
Mar	28.5	17.4	20	4	30.8	19.4	14.1	3	29.7	16	20	4	25.8	13.1	26	11	24.2	13.1	26	11
Apr	24.3	14.4	46	8	27.5	18.9	25.9	6	25.1	12.5	21	5	21.2	11.5	75	14	21.3	11.5	75	14
May	21.4	12	124	14	23.8	14.3	30.2	10	20.7	8.3	27	7	19.2	9.6	99	18	18.6	9.6	99	18
Jun	18.8	10.6	183	17	20.7	13.3	106.8	14	17.6	6.4	32	9	17	8	103	19	16.5	8	103	19
Jul	17.9	9.4	174	18	19.5	13.6	93.3	15	16.7	5	26	9	16.4	7.5	125	22	15.8	7.5	125	22
Aug	18.3	9.4	137	17	20	13.7	66.7	13	18.3	5.3	19	7	16.3	7.2	107	21	15.8	7.2	107	21
Sep	19.6	10.2	80	14	22	14.8	31.4	10	21.7	7.5	15	6	17.5	7.6	84	19	17	7.6	84	19
Oct	22	12.1	56	11	24.3	14.2	19.4	7	25.8	10.9	14	4	19.2	9	88	16	18.7	9	88	16
Nov	24.8	14.4	21	6	26.9	18.1	10.5	4	29.3	14.2	16	4	21.3	10.6	46	12	20.9	10.6	46	12
Dec	28.2	17.1	14	4	29.3	21.0	5.8	2	32.3	16.9	12	3	25.4	12.5	27	10	23.8	12.5	27	10

Perth, like many cities in the world, has lost much of its original vegetation but there are still areas where the flora can be seen and we have selected a few of these locations that are worth visiting.

Near the centre of Perth is Kings Park [1], the largest park set within the confines of a major city in Australia. Here one can walk the many trails that meander through the Banksia - Eucalypt woodland such as **Slender Banksia (Banksia attenuata)**, **Menzies Banksia (Banksia menziesii)**, **Jarrah (Eucalyptus marginata)** and **Marri (Corymbia calophylla)** trees. In spring **Mangles Kangaroo Paws (Anigozanthos manglesii) which is WAs floral emblem,** can be seen through out the park. The map details a section of the park where native plants have been cultivated from all over the south west and it is worth visiting to familiarise yourself with the major plant families. There is always something in bloom throughout the year.

On the west side of the city, close to the coast, is Bold Park [2] which contains examples of coastal limestone frequenting plants as well as **Firewood Banksia (Banksia menziesii)** Tuart (*Eucalyptus gomphocephala*) and some restricted mallees.

The flora of the Darling Scarp is much more prolific. Here, where the Banksia woodland plains meets the laterite and granite outcrops of the Darling Range, is one of the more prolific plant communities in the South West Botanical Province. This section mentions a few areas to visit and gives detailed maps of 5 localities.

Gooseberry Hill [3] drive is well worthwhile in spring, particularly if walking is a problem, as you can slowly drive down, stopping at one or two places on the way. Access is via the suburb of Kalamunda using Williams road. Lascelles Parade and then the scenic Zig Zag Drive (oneway only from the top of the scarp down).

Bickley Brook [4] has a wonderful walk. At times it is hard to imagine that the city is so close. You can start at the bottom at the end of Hardinge Road near Bickley Brook Reservoir in the suburb of Orange Grove and walk up the Kattamorda Heritage Trail following the Bickley Brook valley, or start at Masonmill Road in the suburb of Canning Mills. It is difficult to see the start of the trail at the eastern end but you will see a cleared parking area in the woodland. The trail is 4.5 km one way but you can do part of the walk either way. It is a beautiful walk and you may see **Fuchsia Grevillea (Grevillea bipinnatifida)**, **Wiry Wattle (Acacia extensa)**, **Crinkle leaf Poison (Gastrolobium villosum)**, **Parrot Bush (Dryandra sessilis)**, **Devils Pins (Hovea pungens)** and **Candle Cranberry (Astroloma foliosum)**.

Ellis Brook [5] is another great wildflower area of the Darling Scarp being extremely rich in its flora diversity, particularly on the quartzite and granite slopes. If you are very keen on wildflowers and time is at a premium then visit this area first, ideally between May and December with the high point being September or early October. Walk the granite slopes marked 'Q' opposite the toilets and parking bay. Here you can find the yellow form of the **Many flowered Honeysuckle (Lambertia multiflora)**, **Granite Petrophile (Petrophile biloba)** and **Lemon-scented Darwinia (Darwinia citriodora)**. Return to the parking bay and drive further up the valley to where the road stops and walk up to the falls where there is a lovely spot at the top of the falls to have a picnic. A small back-pack will free your hands for safe walking.

Other areas to visit are Kalamunda National Park [6] where there are many fine walking trails, and John Forrest National Park [7] is good if you want more organised facilities such as a cafe and swimming area.

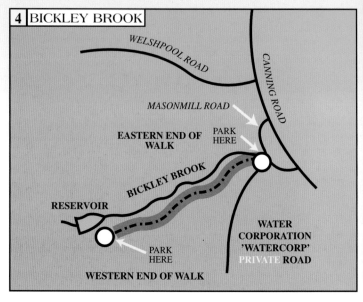

4 | BICKLEY BROOK

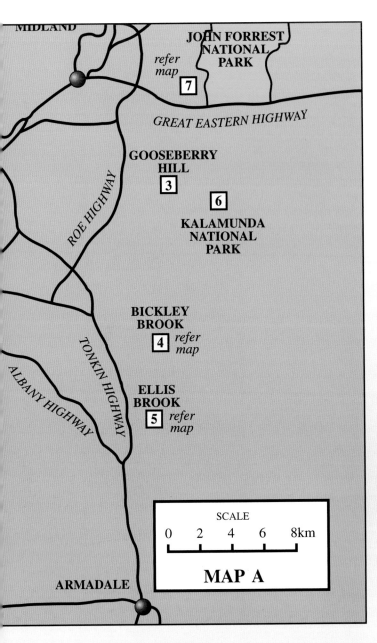

1 KINGS PARK

A KIOSK
B RESTAURANT
P PARKING
T TOILETS

KINGS PARK ROAD
MAIN ENTRANCE
PERTH
MAY DRIVE
FRASER AVE.
NATURAL BUSH
B
P
A
WAR MEMORIAL
BROADWALK
FOREST DRIVE
FOUNTAIN WATERFALLS
OBSERVATION TOWER
NATURAL BUSH
SWAN RIVER
WESTERN AUSTRALIAN NATIVE PLANTS
N

MIDLAND
JOHN FORREST NATIONAL PARK
refer map 7
GREAT EASTERN HIGHWAY
GOOSEBERRY HILL 3
6
KALAMUNDA NATIONAL PARK
ROE HIGHWAY
BICKLEY BROOK 4 *refer map*
ELLIS BROOK 5 *refer map*
ALBANY HIGHWAY
TONKIN HIGHWAY
ARMADALE

SCALE
0 2 4 6 8km

MAP A

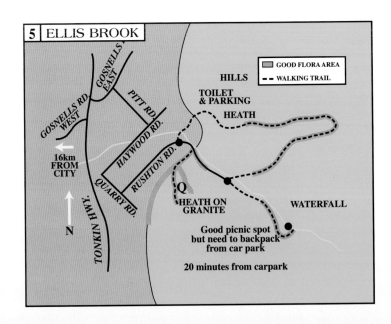

5 ELLIS BROOK

GOOD FLORA AREA
WALKING TRAIL

HILLS
TOILET & PARKING
HEATH
GOSNELLS RD WEST
GOSNELLS EAST
PITT RD.
HAYWOOD RD.
RUSHTON RD.
QUARRY RD.
TONKIN HWY.
16km FROM CITY
N
Q
HEATH ON GRANITE
WATERFALL
Good picnic spot but need to backpack from car park
20 minutes from carpark

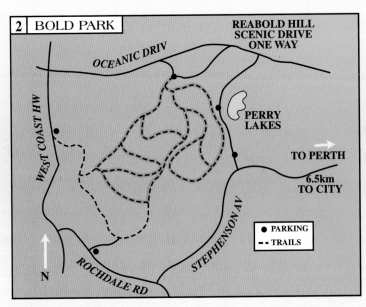

2 BOLD PARK

REABOLD HILL SCENIC DRIVE ONE WAY
OCEANIC DRIV
WEST COAST HW
PERRY LAKES
TO PERTH
6.5km TO CITY
STEPHENSON AV
N
ROCHDALE RD
PARKING
TRAILS

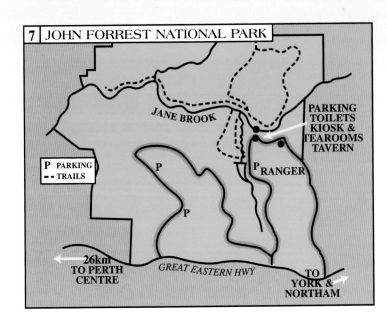

7 JOHN FORREST NATIONAL PARK

JANE BROOK
PARKING TOILETS KIOSK & TEAROOMS TAVERN
P
P
P
P RANGER
P PARKING
TRAILS
26km TO PERTH CENTRE
GREAT EASTERN HWY
TO YORK & NORTHAM

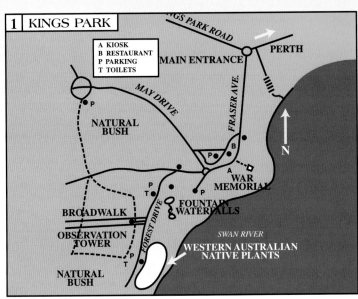

This section covers parks and reserves that can be visited in one day, leaving Perth in the early morning and returning that evening.

A visit to WA will often include the Pinnacles and there are many stops that can be made en route. The first detour can be into Gingin [1] to visit an unusual spot - the local cemetery. In spring you can see a marvellous display of **Kangaroo Paws**, including **Common Catspaw** (*Anigozanthos humilis*) and **Mangles Kangaroo Paw** (*Anigozanthos manglesii*).

Then you can travel back to the highway and 8 kilometres south of the Moore River [2] on the west side. Just past Red Gully Road is a parking bay with a flora road sign located on the edge of Moore River National Park. Here between May and August you will find the restricted but locally common **Rose- fruited Banksia** (*Banksia laricina*) as well as **Winter Bell** (*Blancoa canescens*) and **Purple Tassels** (*Sowerbaea laxiflora*). From mid November to December **Orange Morrisson** (*Verticordia nitens*) is in full bloom below the **Slender Banksia** (*Banksia attenuata*) and **Christmas Tree** (*Nuytsia floribunda*). Both in bloom in December.

Further up the Brand Highway approximately 35 kilometres on the right is Yandan Road leading to Yandan Hill [3]. Here **Mottlecah** (*Eucalyptus macrocarpa*) is common on the road sides, having the largest flowers of any eucalypt. Drive up to the top of Yandan Hill to the picnic parking bay and walk to the edge of the scarp and you will have a splendid view of the Banksia Woodland Plain below you, and around your feet in the height of the season will be a marvellous array of flowers.

Still travelling north you enter Badgingarra National Park [4]. This is one of the state's richest wildflower areas. **Staghorn Bush** (*Daviesia epiphylla*), **Yellow Kangaroo Paw** (*Anigozanthos pulcherrimus*) and **Black Kangaroo Paw** (*Macropidia fuliginosa*), **Scarlet Feather Flower** (*Verticordia grandis*), **Acorn Banksis** (*Banksia prionotes*), **Propeller Banksia** (*Banksia candolleana*) and *Banksia Grossa*, **Summer Coppercups** (*Pileanthus filifolius*) and many of the **Smokebushes** (*Conospermum*) can be found here.

There are several good picnic stops, one just a few kilometres before Cervantes [5]. It is set amongst the shade of some large **Tuart trees** (*Eucalyptus gomphocephala*).

There are shops in Cervantes to obtain wildflower information including some seeds. The Pinnacles [6] are a few kilometres south of Cervantes. You may wish to stay in Cervantes but if returning to Perth, return the way you have travelled. It will be a long day. Remember Cervantes is 243 kilometres from Perth. If by the remote chance you have seen the Pinnacles by mid day, you could return to the Brand Highway and go to Badgingarra, then cut across to Moora [7] and then south stopping briefly in the character monastery town of New Norcia, [8] and then back to Perth. That will certainly be a long day.

Located on the Perth Environs map are some good reserves and areas to visit and one would really need a day in most of them. The Wongan Hills [9] area has much to offer and contains some very rare flora. The map on page 57 will assist you with areas to visit and but you should really spend an overnight in Wongan to appreciate the various granite outcrops and sandplains as well as the hills themselves. Reynolds Reserve not far north of Wongan, is noted for its magnificent Verticordia display in November to early December. Just north of the town of Piawaning [10] is a very rich flora drive. One of the loveliest flora drives in W.A. is just south of New Norcia, called the Old Plains Road [11]. Here particularly from September to October can be seen a fine example of road side wildflowers which demonstrates how important it is to conserve our precious flora road verges. At the junction of Old Plains Road and Calingiri Road is another small reserve [12] named in honour of one of WA's eminent botanists, Rica Erickson; here you can find **Woolly Bush (Adenanthos cygnorum)**, **Pingle (Dryandra carduacea)**, **Marble Hakea (Hakea incrassata)**, **Fox Banksia (Banksia sphaerocarpa)** and **Roadside Tea-tree (Leptospermum erubescens)**.

Wongamine Reserve [13] located off the Goomalling-Toodyay Road between Forest and Bejoording Road contains a few patches of rich Kwongan above the laterite hills, as well as **Wandoo (Eucalyptus wandoo)**, **Powder bark Wandoo (Eucalyptus accedens)**, **Brown Mallet (Eucalyptus astringens)** and **York Gum (Eucalyptus loxophleba)**. In this one small reserve alone there are 24 species of orchid, again showing how we must preserve our remnant island reserves from further land clearing.

Further south on the Brookton Highway is Boulder Rock [14], a good example of a large granite outcrop. Here typical granite wildflowers can be seen such as **Pincushion Plant (Borya sphaerocephala)**, **Mouse Ears (Calothamnus rupestris)** and **Trigger plants (Stylidium sp.)** and **Sundews (Drosera sp)**. An unsealed road takes you up to Mount Dale [15], giving superb views across the **Jarrah (Eucalyptus marginata)** and **Marri (Corymbia calophylla)** Forest. Before the peak is a picnic spot with interesting granite flora like **Sea Urchin Hakea (Hakea petiolaris)**.

Christmas Tree Well [16] as the name implies has **Christmas Trees (Nuytsia floribunda)** and an old stage coach well as well as some some picnic tables.

Just a few kilometres further, on the north side of Brookton Highway, is Yarra Road, on some maps Jarrah road. This road cuts through to the Perth - York Road. Travelling along this dirt road you can turn due east down Qualen Road towards the Wandoo Conservation Park [17]; along the many tracks that criss cross this area you will find pockets of sandheath and beautiful open stands of **Wandoo (Eucalyptus wandoo)** and **Powder bark Wandoo (Eucalyptus accedens)** trees. It is hard to imagine that Perth is only an hour away, so we are privileged to have this fine woodland so close to Perth. It is recommended that adequate maps are taken with you, as even though this country is close to Perth there are few people who venture into this woodland. The area is especially noted for its rare orchids as well as a rare Fabaceae (pea plants) Synaphea and a Allocasuarina sp. A drive to the old character towns of York and Toodyay can be combined with visiting these areas.

Two reserves not far from each other are well worth visiting. Dryandra Nature Reserve [18] and Boyagin Rock Nature Reserve [19]. Both (refer map page 52 and 53) contain large stands of open Wandoo Forest. Within Boyagin Reserve are two very large granite rocks, one of them having fine stands of **Ceaesia (Eucalyptus caesia)**. In the sheoak woodlands below the rock in early spring can be found many species of orchid that thrive in the high acidic soils below the **Rock Sheoak (Allocasuarina huegeliana)**.

Dryandra Forest is a unique reserve not only having some good Kwongan heath but containing some unusual marsupials such as Numbat and Tammar Wallaby, Brush tailed Bettong and Echidna, so there is much to see here.

The journeys to the Pinnacles, Wongan Hills, Boyagin Rock or Dryandra can be achieved in a day but really to enjoy them properly you should stay overnight.

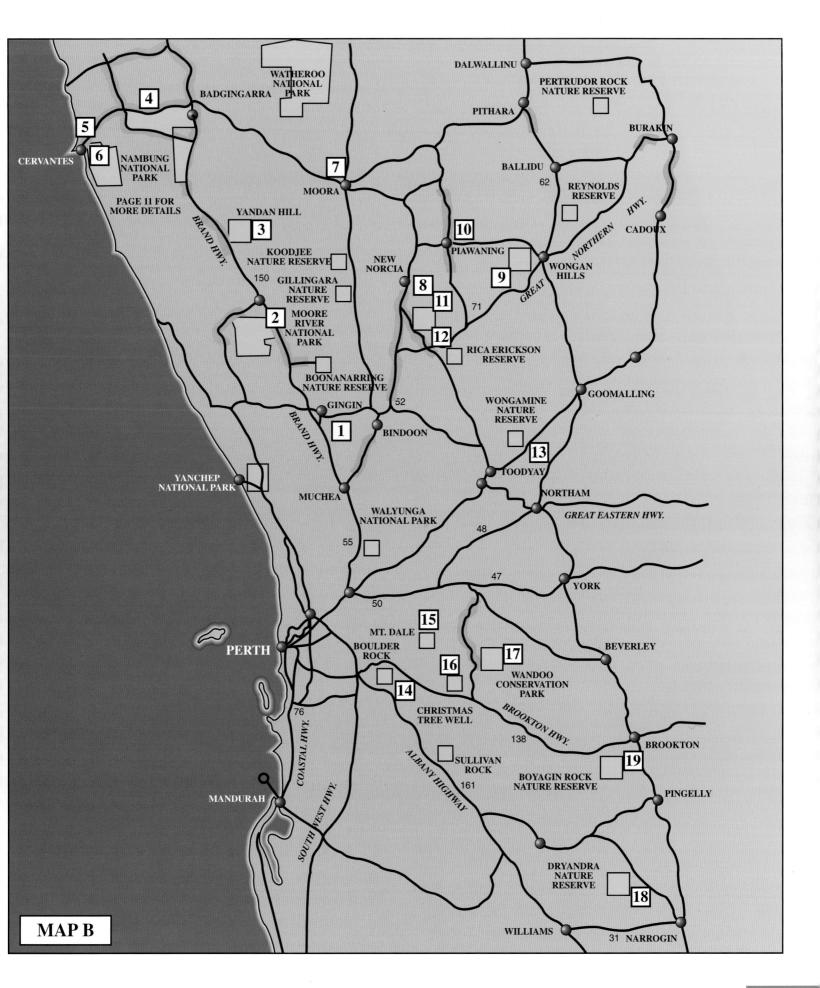

MAP B

There is much to see travelling north and it may be best to do this prior to travelling south, as the Spring flowering season start earlier in the north. If travelling up the Brand Highway, then many of the locations mentioned in the Perth Environs section could be visited. Around the Badgingarra area are several national parks and reserves. They contain some of the highest plant diversities in the south west.

Mt Lesueur 1 is one of them and after the Stirling Range and the Fitzgerald River National Park in the south, is one of the most important conservation reserves we have, with over 800 plant taxa known including 111 regionally endemic plants. The area is dominated by laterite hills or mesas. Here you can see **Honey Bush** (*Hakea lissocarpha*) as well as rare plants like **Cork Mallee** (*Eucalyptus suberea*) and **Lesueur Hakea** (*Hakea megalosperma*).

Coomallo Creek 2 picnic area on the Brand Highway has a lovely stand of Powderbark Wandoo as well as kwongan heath and is easily accessed from the parking area. Travelling along the Jurien Road to Mt Lesueur there are two very good picnic spots less frequented than the highway one at 11.5 and 17.5 km mark from the highway.

East and north of Mt Lesueur are two very similar reserves, Alexander Morrison 3 and Tathra National Park 4 ;both have extensive Kwongan heath with **Shaggy Dryandra** (*Dryandra speciosa*), **Fringed Bell** (*Darwinia nieldiana*) and **Fishbone Banksia** (*Banksia chamaephyton*).

The Brand Highway roadside verges contain a bewildering number of wildflowers, particularly north and south of Eneabba 5 .Around the shores of Lake Indoon 6 on the Leeman road the **Elegant Banksia (Banksia elegans)** is found along with the more prolific **Hookers Banksia (Banksia hookeriana)**. To differentiate **Hookers** from **Acorn Banksia, Orange Woolly Banksia** and **Burdetts Banksia** compare the leaf structures (refer map page 84). About 35 km south of Dongara the northern heath and mallee shrublands give way to coastal Wattle country and farming country. It is not until north of Northhampton that the sandplains reappear showing the higher species diversity, although the local hills of Geraldton area contain a very localised and interesting flora.

Turning off the highway near Ajana brings you into the Kalbarri National Park 7 (see map page 93) with over 180,000 hectares consisting mostly of open Kwongan, with a wealth of wildflowers. In mid October along the roadsides, **White Plume Grevilleas** (*Grevillea leucopteris*) line the way and later in December **Sceptre Banksia** (*Banksia sceptrum*) and then **Orange Woolly Banksia**(*Banksia victoriae)* take over the flower show with the bright pink **Woolly Featherflowers** (*Verticordia monadelpha*).

The gorges of the Murchison River are not only worth visiting for their scenic value but also there are some plants that frequent the gorge slopes not seen on the heath.

Travelling north again up to, Monkey Mia, 8 you will pass a few patches of rich kwongan heath north of the Murchison River containing **Red Pokers (Hakea buccuulenta)**, **Ashby's Banksia** (*Banksia ashbyi*) and **Prickly Plume Grevillea** (*Grevillea annulifera*).

The transitional woodland and mulga begin to appear and it's here in early spring that the everlastings prolferate.

Returning to Geraldton and driving out to Mullewa 9 you will pass by more Kwongan heath. From late August to mid September the unique **Wreath Leschenaultia (Lechenaultia macrantha)** is in bloom. There is a gravel pit near Pindar 10 where you should see them but ask the local people in Mullewa for directions or if travelling south to Morawa look for signs to the Gutha cemetery 11; as they are also there.

For those venturing out into the mulga zone towards Yalgoo and Mount Magnet to view the everlastings, make sure you do carry enough fuel and water. The road from Mullewa to Mount Magnet is sealed, (conventional vehicles are fine), as is the highway back down from Mount Magnet to Wubin.

Rain falls infrequently in this area so the everlastings will be reliant on good winter rains for a good display, but when they do it is well worth the visit. In this acacia open woodland the many types of **Poverty Bush (Eremophila species)** come into their own as well as the many species of everlastings and mulla mulla. Plants to see are **Splendid Everlasting (Rhodanthe chlorocephala** subsp. *splendida*), **Sticky Everlasting (Lawrencia davenportii)**, **Tall Mulla Mulla (Ptilotus exaltatus)**, **Native Fuchsia (Eremophila maculta)** and **Kopi Poverty Bush (Eremophila miniata)**.

There is accommodation in most of the outback towns, sometimes basic but usually always friendly. An excellent guide for accommodation in WA, including farm stays, is the RAC 'Touring and Accommodation Guide'

Travelling back down the Great Northern Highway you pass by Paynes Find and then re enter the transitional woodland and again encounter the species rich northern heaths before Wubin.

Between Mullewa and Wubin there are some good roadside verges although the best are from just south of Wubin to just south of Perenjori 12 , so if you are not travelling east from Mullewa to the mulga country you can head south to Wubin or Mingenew.

Another productive area is the central Northern Region with much to see, such as Coalseam Gorge 13 near Mingenew and the drive from Mingenew all the way down to Moora.

If returning to Perth from Wubin and time permits, it is more rewarding to travel to Ballidu and Wongan Hills then cross to Perenjori travelling 5 km north to see a rich area of roadside verge and then drive west to Waddington and south to New Norcia 14. The roadsides south of New Norcia from approximately 7 km south to 35 km south are quite rich in wildflowers in spring and some unique species occur down this road. Small reserves like Udumung, Burroloo Well, Seven Mile Well and Barraca Reserve between New Norcia and Bindoon have a good selection of plants, particularly Udumung, containing **Common Brown Pea (Bossiaea eriocarpa)**, **One sided Bottlebrush (Calothamnus quadrifidus)**, **Wavy-leaved Hakea (Hakea undulata)** and other species.

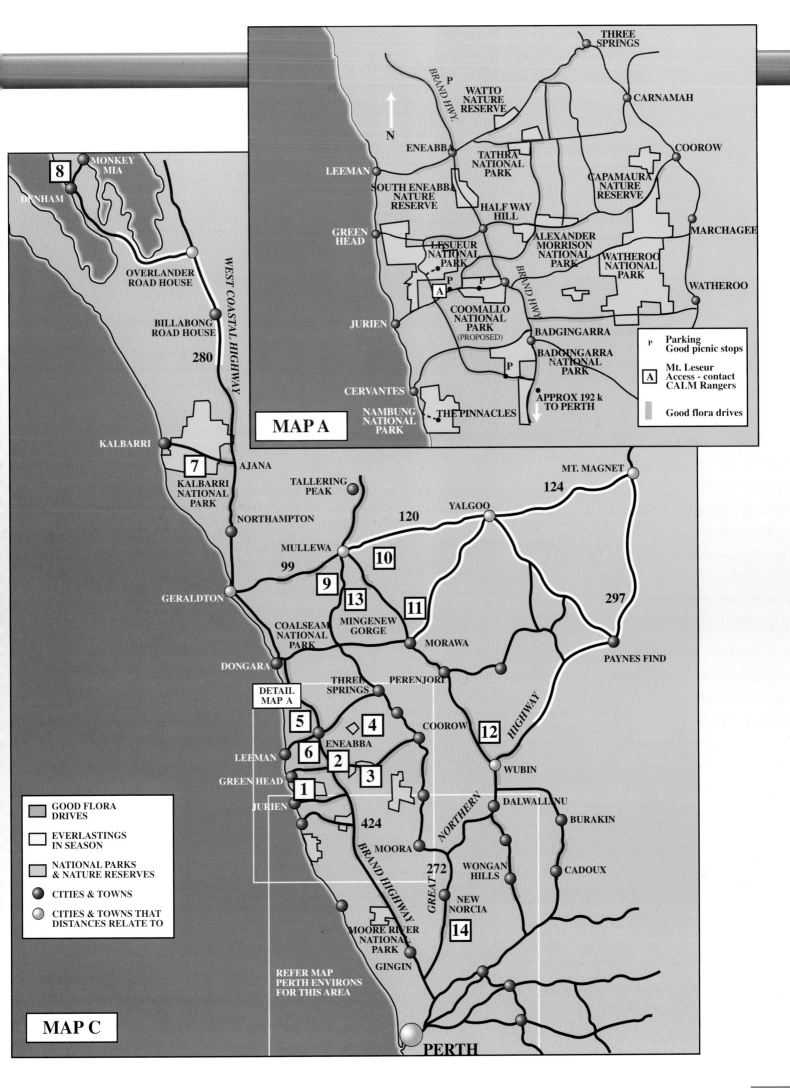

MAP A

THREE SPRINGS

BRAND HWY.

WATTO NATURE RESERVE

CARNAMAH

ENEABBA

TATHRA NATIONAL PARK

COOROW

LEEMAN

SOUTH ENEABBA NATURE RESERVE

HALF WAY HILL

CAPAMAURA NATURE RESERVE

MARCHAGEE

GREEN HEAD

LESUEUR NATIONAL PARK

ALEXANDER MORRISON NATIONAL PARK

WATHEROO NATIONAL PARK

WATHEROO

BRAND HWY

A

JURIEN

COOMALLO NATIONAL PARK (PROPOSED)

BADGINGARRA

BADGINGARRA NATIONAL PARK

CERVANTES

APPROX 192 k TO PERTH

NAMBUNG NATIONAL PARK

THE PINNACLES

P Parking
Good picnic stops

A Mt. Leseur
Access - contact CALM Rangers

Good flora drives

MAP C

8

MONKEY MIA

DENHAM

OVERLANDER ROAD HOUSE

WEST COASTAL HIGHWAY

BILLABONG ROAD HOUSE

280

KALBARRI

7

KALBARRI NATIONAL PARK

AJANA

TALLERING PEAK

MT. MAGNET

124

NORTHAMPTON

YALGOO

120

MULLEWA

99

10

GERALDTON

9

13

11

COALSEAM NATIONAL PARK

MINGENEW GORGE

MORAWA

297

PAYNES FIND

DONGARA

THREE SPRINGS

PERENJORI

DETAIL MAP A

5

4

6

ENEABBA

2

COOROW

12

HIGHWAY

LEEMAN

3

GREEN HEAD

1

WUBIN

JURIEN

424

DALWALLINU

BURAKIN

MOORA

NORTHERN

272

WONGAN HILLS

CADOUX

BRAND HIGHWAY

GREAT

NEW NORCIA

14

MOORE RIVER NATIONAL PARK

GINGIN

REFER MAP PERTH ENVIRONS FOR THIS AREA

GOOD FLORA DRIVES

EVERLASTINGS IN SEASON

NATIONAL PARKS & NATURE RESERVES

CITIES & TOWNS

CITIES & TOWNS THAT DISTANCES RELATE TO

PERTH

11

There are many routes to take leaving Perth to enter the south west region. If you are travelling down to Busselton, Margaret River or as far as Denmark, either take the coast road or the South Western Highway to Bunbury, and then make your decision which way you intend to travel.

Before Busselton is the small Tuart Forest National Park [1] containing some magnificent **Tuart trees (Eucalyptus gomphocephala)**.

Sugar Loaf Rock [2] near Dunsborough has good coastal heath and a picturesque coast line and Caves Road leads further south, a pleasant drive that will take you to Boranup Karri Forest [3] which is 100 year old regrowth forest. You can turn at Conto Road and travel through the forest south coming back on the main road to Augusta. The **Karri (Eucalyptus diversicolor)** forest does not have the flora diversity of the Kwongan but still has some spectacular plants such as **Peppermint (Agonis flexuosa)**, **Tree Hovea (Hovea elliptica)**, **Chorilaena (Chorilaena)**, **Tassel Flower (Leucopogon verticillatus)**, **Karri Hazel (Trymalium spathulatum)**, **Coral Vine (Kennedia coccinea)** and **Cut leaf Hibbertia (Hibbertia cuneiformis)**.

One of the richest flora areas in the deep south west is along Scott River Road [4] containing many rare and restricted plants, but unless you are are a very keen plant enthusiast, you may wish to keep driving towards Pemberton and Walpole. Scott River Road will take you well out of your way down dirt roads. The country does not have such splendid scenery as the tall Karri forest, but many species may be found in the swamps.

Around Pemberton in the Beedalup, Warren and Brockman National Parks [5] are some of the finest old growth Karri forest drives.

If time permits on your way to Walpole take a detour off the Manjimup - Walpole road 85 km south from Manjimup, turn left down Beardmore Road. You will pass Fernhook Falls at the 5.5 km mark to a lovely picnic spot on the Deep River. Travel on over Thomson Road and drive on up to the base of Mt Frankland [6]. Here is a good picnic spot and a walk to the granite summit, but I don't recommend the walk for everyone, as there are two steep steel ladders built on the walk to aid climbing (be careful). Here you can find 60cm tall leek orchids in late spring and granite-frequenting verticordias like **Plumed Featherflower (Verticordia plumosa)** and **Free-flowering Lasiopetalum (Lasiopetalum floribundum)**. The views from the summit give a wonderful panoramic vista and the extent of the south west forests. Leaving here you can return to the North Walpole road and head south to Walpole.

If you wish to see **Red Flowering Gum (Corymbia ficifolia)** in full bloom in its natural habitat drive along Ficifolia Road [7] named after this plant, but you must be there in December to January to see the beautiful red flowers. Also a visit to the Tree Top Walk will give you a bird's eye view of the tall **Tingle (Eucalyptus jacksonii)** and Karri **(Eucalyptus diversicolor)** .

After the towns of Walpole and Denmark you will travel towards Albany. Past the village of Elleker [8] turn right down Grassmere Road and you will find **Red Swamp Banksia (Banksia occidentalis)** on the right hand side. It flowers in December and February. Back on the Albany road, along the roadside verges you may see **Albany Bottlebrush (Callistemon speciosa)** and **Swamp Bottlebrush (Beaufortia sparsa)**. You can see why it is useful to know the latin names as they are each of a different genus.

Now you have reached Albany with its magnificent harbour and the rugged peninsular park of Torndirrup National Park [9] .Here, besides the many tourist lookouts, you can see at Stony Hill fine examples of **Granite Banksia (Banksia verticillata)** and then at Salmon Holes is also **Cut - leaf Banksia (Banksia praemorsa)**, both restricted to this southern coastline. There is much to see around Albany and travelling north you will come across the granite hills of the Porongurup National Park [10] with another lovely walk up through tall karri trees. Here you may come across the rare **Mountain Villarsia (Villarsia marchantii)**.

North along Chester Pass Road is the magnificent Stirling Range National Park [11] (Refer map page 64). You are now in an area of extremely high plant diversity with over 1500 species in this park alone and many of the plants exist only here amongst the quartzite mountains of the Stirling Range. These endemics include **Mountain Pea (Nemcia leakiana)**, **Giant Andersonia (Andersonia axillifolia)**, **Stirling Range Bottlebrush (Beaufortia heterophylla)**, **Stirling Range Banksia (Banksia solandri)**, **Mountain Kunzea (Kunzea montana)**, **Isopogon latifolius** and the unique **Mountain Bells (Darwinias)** found on different peaks and valleys within the park.

The lower slopes of the ranges are at their best from the beginning of September to the end of October and then the flowering period drops off quickly either side of these months.

If you are travelling to Albany from Perth via the Albany Highway, Sullivans Rock [12] is worth walking up onto the lower slopes of the granite outcrop, opposite the parking bay. Further south is Mary Pool [13] just past the turn off to Woodanilling. Here amongst the sheoak woodland near the creek in September and October you should find several orchids such as **Common Donkey Orchid (Diuris corymbosa)**, **Cowslip Orchid (Caladenia flava)**, **Pink Enamel Orchid (Elythranthera emarginata)**, and **Blue Fairy Orchid (Cyanicula deformis)**.

You can turn off to the Stirling Range National Park (refer map 64) at Cranbrook [14].The local people will tell you where there is a wildflower walk on the edge of town, which can be very productive in spring, particularly for orchids. They also have a wonderful wildflower show in spring, as do the towns of Albany, Kojonup, Ongerup and Ravensthorpe. Most plants are identified and the display is staffed by very knowledgeable volunteers, all too ready to help those interested in wildflowers.

We have mentioned Dryandra and Boygin Rock Reserves as two very good places to visit on the Perth Environs Map, particularly Dryandra, but you will note more reserves marked on the South West Region map. There are sketch maps of Charles Gardner Nature Reserve [16] and Boolanooling Nature [15] Reserve (page 105) and Tarin Rock Reserve page (74) [17] Harrismith (page 61) [18] .All can be visited on a general route to Esperance. These reserves are a botanist's delight. They are not spectacular mountain ranges but low undulating kwongan heath lands and are extremely rich in their flora containing some restricted species.

Two other very special reserves are Tutanning [19] and Dongolocking Nature Reserves, [20] again containing very interesting plants and varied landscapes. But they are very difficult to access and really should only be visited after seeking advice from the local Departement of Conservation and Land Management Offices in Katanning or Narrogin. They can assist you where to go.

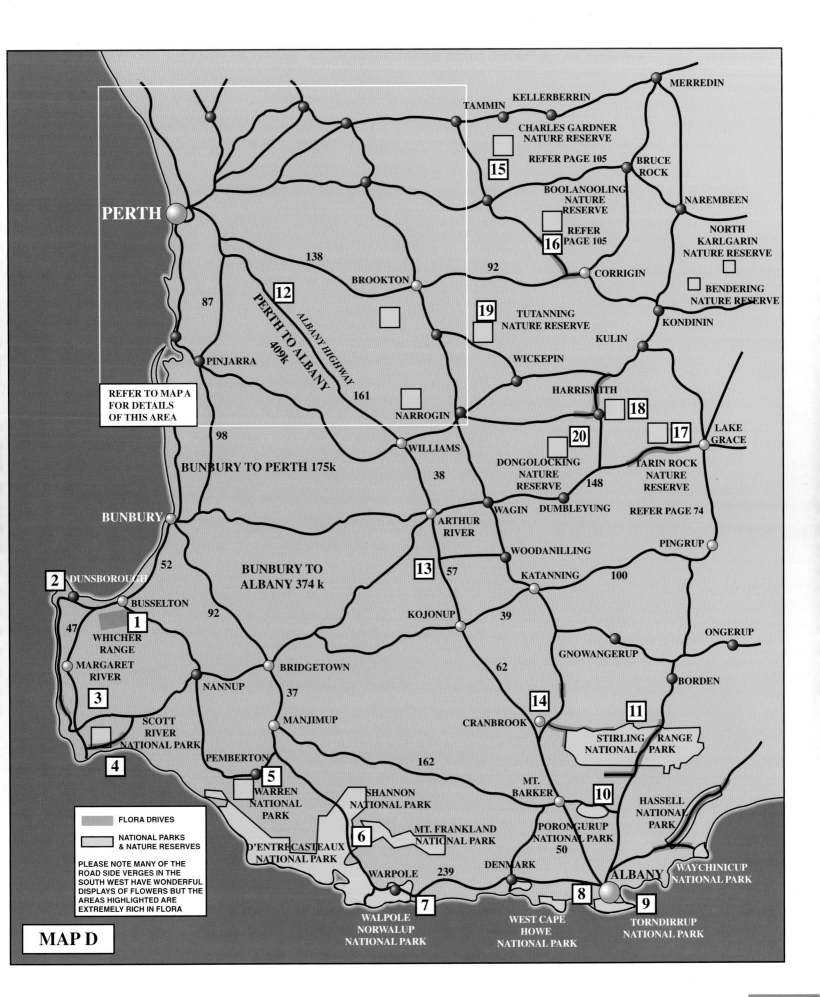

MERREDIN

KELLERBERRIN

TAMMIN

CHARLES GARDNER
NATURE RESERVE

15 REFER PAGE 105

BRUCE
ROCK

NAREMBEEN

BOOLANOOLING
NATURE
RESERVE

NORTH
KARLGARIN
NATURE RESERVE

16 REFER
PAGE 105

138

92

BENDERING
NATURE RESERVE

PERTH

BROOKTON

CORRIGIN

KONDININ

87

12

KULIN

19 TUTANNING
NATURE RESERVE

PINJARRA

WICKEPIN

HARRISMITH

REFER TO MAP A
FOR DETAILS
OF THIS AREA

161

NARROGIN

18

17 LAKE
GRACE

98

WILLIAMS

20

20

DONGOLOCKING
NATURE
RESERVE

TARIN ROCK
NATURE
RESERVE

BUNBURY TO PERTH 175k

38

148

REFER PAGE 74

BUNBURY

WAGIN DUMBLEYUNG

PINGRUP

ARTHUR
RIVER

WOODANILLING

52

BUNBURY TO
ALBANY 374 k

13 57

KATANNING

100

2 DUNSBOROUGH

KOJONUP

39

ONGERUP

BUSSELTON

92

47

1

WHICHER
RANGE

GNOWANGERUP

62

MARGARET
RIVER

BRIDGETOWN

BORDEN

3

NANNUP

37

14

CRANBROOK

11

SCOTT
RIVER
NATIONAL PARK

MANJIMUP

STIRLING   RANGE
NATIONAL   PARK

4

PEMBERTON

162

MT.
BARKER

10

HASSELL
NATIONAL
PARK

5 WARREN
NATIONAL
PARK

SHANNON
NATIONAL PARK

PORONGURUP
NATIONAL PARK

FLORA DRIVES

6 MT. FRANKLAND
NATIONAL PARK

50

NATIONAL PARKS
& NATURE RESERVES

D'ENTRECASTEAUX
NATIONAL PARK

DENMARK

WAYCHINICUP
NATIONAL PARK

PLEASE NOTE MANY OF THE
ROAD SIDE VERGES IN THE
SOUTH WEST HAVE WONDERFUL
DISPLAYS OF FLOWERS BUT THE
AREAS HIGHLIGHTED ARE
EXTREMELY RICH IN FLORA

WARPOLE

239

8 ALBANY

9 TORNDIRRUP
NATIONAL PARK

7

MAP D

WALPOLE
NORWALUP
NATIONAL PARK

WEST CAPE
HOWE
NATIONAL PARK

Part of this map is covered by the Perth Environs map and the South West Region map, so you will already have information for the areas between Perth and Lake Grace.

Many visitors to Western Australia want to go to Wave Rock [1] near Hyden. It is a full day but there are interesting flora places en route to visit. If you are staying at Hyden you will have time to visit a very good reserve mentioned before, Boyagin Rock [2] (refer map page 52). The turn-off is just 18 km before Brookton on the Brookton Highway. Turn right travelling south on the old York - Williams road and at the 9 km mark you will see a sign pointing left to Boyagin Rock (refer map page 52); travel along this small dirt road (2 wheel drive OK) entering the main reserve and you will pass a small patch of heath. See how rich this small area is with **Showy Dryandra (Dryandra formosa)**, **Drummonds Gum (Eucalyptus drummondii)** and an endemic *Synaphea sp.* Further into the reserve you see fields on both sides; take the turning marked Boyagin Rock along side a field and you will come to a small picnic area. Park here and walk up one of the granite outcrops.

When you leave, go back to the track with the Boyagin Rock sign and this time turn left (due east) between the fields and the reserve will start again. Keep going through the reserve and exit on the eastern side. Turn left on Walwalling Rd. travelling to the Great Southern Highway, then turn north to Brookton; then you are back on the road to Hyden in the town of Brookton.

About 18 km before Corrigin there is a small patch of kwongan containing many plant species. Just turn north up Jubuk Road [3] and park anywhere off the road and walk the heath; there are several Verticordias in late spring including **Common Cauliflower (Verticordia eriocephala)** and **Painted Featherflower (Verticordia picta)**.

If time permits travel a little way up the old fence road called the Corrigin - Quairading Road. The turn off is only 11 km past the last stop and goes north. About 4 km along this road you will start to see some good roadside verges surviving between the cleared farmlands. The junction of Lohoar and Gill Road [4] is particularly good and the next 6 km or so is still prolific. If you were on long day trip to just this area you could visit Boolanoolling Reserve [5] , a little known but very productive reserve (Refer map page 114) but you will need accurate maps to find this reserve.

If you wanted a quiet spot to picnic then return to the Brookton - Corrigin Road and go straight on down the Dudinin Road [6] for 6 km. Cross over the railway line and turn immediately left. Right on this corner are some remnant patches of **Flat-leaved Wattle (Acacia glaucoptera)** and **Sand Mallee (Eucalyptus eremophila)**. A little way down park under the shade of the **Wandoo** and **York Gums (Eucalyptus loxophleba)** (please do not light fires here). Then return to the Brookton - Corrigin Road. Just 2 km before Corrigin is a drive up to a look out surrounded by a water catchment reserve where more plants can be seen.

After the township of Kondinin the main road passes through Karlgarin Reserve [7] with fine stands of mallee trees and **Salmon Gum (Eucalyptus salmonophloia)**. The large reserves of West Bendering and Bendering [8] due north west of Karlgarin Reserve are very extensive and contain some very interesting plants in the kwongan areas as well as very diverse eucalypts, particularly the mallees. You will see many species here. Don't forget to take those eucalypt reference books with you.

Opposite Wave Rock out of Hyden there is a very good shop with a fine display of dried wildflowers. Most Western Australians will know that there are many wonderful granite rock outcrops to visit. Around Hyden

there are The Humps, Graham Rock, Emu Rock, Varley Rocks and Holt Rock and south of Lake King are two fine granite outcrops called Mt Madden and Pallarup Rocks, both with good picnic sites.

There are some very specialised plants that grow only on or in the shallow soils adjacent to these granite outcrops, such as **Granite Kunzea (Kunzea pulchella)**, **One sided cBottlebrush(Calothamnus quadrifidus)**, **Baxters Kunzea (Kunzea baxteri)**, and **Caesia (Eucalyptus caesia)**, and south of Hyden is the uncommon *Grevillea magnifica sub sp. remota* **Pink Pokers**, very similar to (*Grevillea petrophiloides*).

Between Hyden and Lake King in spring, some of the roadside verges are a mass of colour. Travelling south before Lake King is a reserve called Kathleen [9] . Just before this on the west side of the road is a gravel pit reserve. Here you can find 5 separate **featherflowers (Verticordias)** as well as **Flame Grevillea (Grevillea eriostachya)**, **Curly Grevillea (Grevillea eryngioides)** at its southern limit and **Lemanns Banksia (Banksia lemanniana)** at its northern limit.

The more adventurous traveller can take the dirt road due east of Lake King to Frank Hann [10] and Peak Charles National Park [11] , but you must have adequate fuel, food and water supplies and ideally a 4WD vehicle. You must also be careful as these roads can become impassable after heavy rains. If you do visit this area you will see some of the most extensive semi-arid woodlands in Australia. It would be wonderful if more of it could be set aside as reserve as those who know this country realise what a wealth of flora exists here.

If you take the road to Cascades [12] south of the Lake King-Norseman Road, the first 30 km has very good roadside verges and some years it will be in full bloom as late as the end of November, with such plants as **Red Kangaroo Paw (Anigozanthos rufus)**, **Red Rod (Eremophila calorhabdos)** and **Fuchsia Gum (Eucalyptus forrestiana)**.

If you head south from Lake King towards Ravensthorpe [13] you will be entering yet another flora-rich zone. North of Ravensthorpe is Floater Road, which will take you up to the hills behind Ravensthorpe. Here you can see **Ravensthorpe Bottlebrush (Beaufortia orbifolia)**, **Ravensthorpe Hakea (Hakea obtusa)**, **Bushy Yate (Eucalyptus lehamannii)**, **Tennis Ball Banksia (Banksia laevigata)** and **Shaggy Dog Dryandra (Dryandra foliosissima)**.

It is a long drive to Esperance from Ravensthorpe and the best roadside verges are at the Ravensthorpe end, but when you arrive in Esperance you will be treated to wonderful ocean views and the scenic coastal drive along Twilight Beach Road [14] is certainly recommended.

Further east is the road to Cape Le Grand [15], Cape Arid and Mt Ragged. Cape Le Grand is the closest National Park, with a picturesque camping beach at Lucky Bay. There are fine examples of coastal heathlands and rugged granite outcrops. Here you can find **Teasel Banksia (Banksia pulchella)**, **Showy Banksia (Banksia speciosa)** and **Creeping Banksia (Banksia repens)**, **Corky Honeymyrtle (Melaleuca suberosa)**, **Silver Tea Tree (Leptospermum sericeum)**, **Chittick (Lambertia inermis)** and **Tallerack (Eucalyptus tetragona)**.

If you still want to travel further east to Cape Arid National Park you will see many of the species just mentioned, along the roadside verges keep your

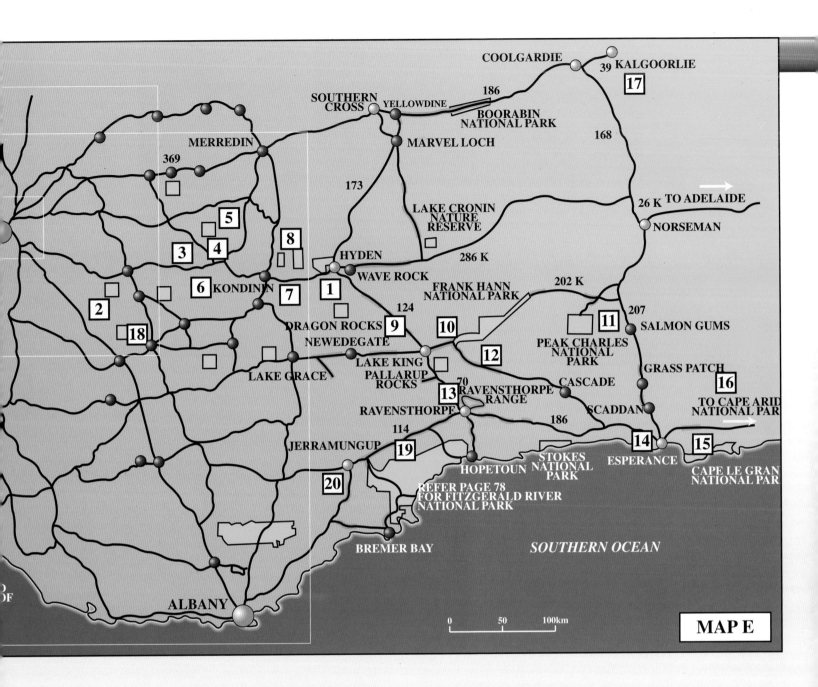

eye out for **Heath Lechenaultia (Lechenaultia tubiflora)**. The track to Mt Ragged [16] is fine for the first 4 km then it becomes a real 4WD track, which is often flooded and is only recommended for experienced 4WD drivers. If you do venture on this track, you will pass two banksias restricted to the Esperance region, *Banksia pilostylis* and a prostrate **banksia *Banksia petiolaris***. On Mt Ragged there are some unique plants as this is an outlying quartzite rock similar to the Barrens and the Stirling Range, surrounded by mallee shrublands.

North of Esperance you can travel to Norseman and Kalgoorlie [17], but this takes you outside the confines of the South West Botanical Province and also outside of the parameters of this book; but do not let this stop you, as there are some wonderful eucalypt woodlands. In the Kalgoorlie area, are more than 60 eucalypt species. If you decide to travel this way on your return to Perth you should not only visit Kalgoorlie and Coolgardie, but on your way back just before Southern Cross, you should stop in the kwongan heaths of Boorabbin National Park.

We started this section with a journey to Hyden and south to Lake King and then on to Esperance, but of course there are many ways to travel south east from Perth and you may wish to travel via Dryandra Forest [18] then on to Narrogin and down the Great Southern Highway to Katanning.

One park that should not be missed if you are travelling further than Albany is the Fitzgerald River National Park [19] one of the botanic wonders of the world. (refer map page 78). From Albany, you drive here via the Great Southern Highway, turning off to Borden and then to Jerramungup. The Fitzgerald River National Park to date has 1883 plant taxa. There are 72 species endemic to this park. It has over 70 orchids and 16 banksias.

Travelling via Jerramungup [20], one can enter the northwest corner of the park. There is an information bay here that has a detailed map of the various tracks where you may travel. As you travel down the Pabelup Track you can see many plants such as **Royal Hakea (*Hakea victoria*)**, **Nodding Banksia (*Banksia nutans*)**, **Southern Plains Banksia (*Banksia media*)**, **Prostrate Banksia (*Banksia gardneri*)**, **Cayleys Banksia (*Banksia caleyi*)**, **Baxters Banksia (*Banksia baxteri*)** and the endemic **Qualup Bell (*Pimelea physodes*)**, named after the Quaalup homestead where you can stay.

The Barren mountains may look barren from a distance but if you visit West Mt Barren or East Mt Barren you will find very restricted plants such as **Mountain Banksia (*Banksia oreophila*)**, **Barrens Regelia (*Regelia velutina*)** and **Dense Clawflower (*Calothamnus pinifolius*)**.

You can exit the park via Bremer Bay or you may wish to travel to the eastern end of the park and a journey down Hamersley Drive will provide more new plants. You will exit the park near Hopetoun where you can drive back to Ravensthorpe, then on to Esperance or back to Albany or Perth.

Chapter II

The Wildflowers of the South West

The term "flora" is often heard or read about when the subject of wildflowers arises and "the flora" means all the plants of a particular area.

The area discussed in this guide, the south west of Western Australia has one of the most significant and diverse floras on the earth and there is a fascinating range of reasons for this, to do with several broad factors:

- ancient Gondwanan relationships with neighbouring continents,

- a very old and stable land surface with nutrient-poor soils,

- varied climates over the millennia,

- an island situation, water to the west and south and deserts to the north and east, and

- comprehensive long term availability of pollinating animals.

Let's look at each of these broad factors and offer some explanation.

1. Ancient Gondwanan relationships with neighbouring continents.

Australia was once part of the southern hemisphere supercontinent Gondwana, along with the other landmasses Africa, Madagascar, India, New Zealand, South America and Antarctica, as well as other smaller fragments such as New Caledonia and Mauritius. Continental drift separated these lands over a period from around 110 million years ago when India separated, to around 65 million years ago when Australia sheared away from Antartica and drifted north to its present location . The lands of Gondwana shared common plants that have evolved into different although related floras, evidence of which is easily seen in the modern floras of all these lands.

Some fine examples are in the Protea family Proteaceae, represented by among others the genus *Protea* and *Leucadendron* in South Africa, *Banksia* and *Telopea* in Australia, *Persoonia* in Australia and New Zealand, *Lomatia* in Australia and temperate South America and *Dryandra* in Western Australia only.

2. An extremely old and stable land surface with nutrient-poor soils.

Little has changed in the basic landform structures over the last 45 million years in what is now the south west of Western Australia. The area is underlain by a huge granite shield known as the Yilgarn Craton.

as well as the areas to the west and south, as coastal plains that are now reminders of what were once huge rift valleys between India to the west and Antarctica to the south.

It is well established that the area is one of the oldest and least changed terrestrial land surfaces on the Earth, having remained above sea level for well over 200 million years.

Volcanic and glacial activity assist greatly in soil renewal by making available nutrients and minerals vital for fertility. The last time any significant volcanic activity occurred in the now south west of Western Australia was 160 million years ago, when basalt rocks were formed as part of the geological stresses in the Earth's crust associated with a weak point in the dynamic edge of the then joined Indian and Australian continental plates. No glacial activity has influenced the area as no conditions suitable have prevailed, such as large valleys in mountain systems to allow glaciers to form and grind out new soil types.

The varied soil types of the south west of Western Australia are consequently extremely old and relatively nutrient-poor having had no renewal for so long. They are strangely complex having had origins in geological and climatic events over an incredibly long history. Climatic conditions have also forced the re-positioning and reworking of these old soils to create the many even more nutrient-poor contemporary soils.

3. Varied climates over the millennia.

The very long time that the area has been exposed as a terrestrial land surface has seen it exposed to many climatic regimes. Evidence of the different climates can be seen in the present day rocks and soil types; for example the red/brown rock type laterite that is seen in the Perth hills and elsewhere can only be formed under tropical or monsoonal conditions. Where very wet and hot conditions allow, chemical actions to leach out the iron or alumina from underlying rocks (parent rocks), creating the spherical or nodular formation of gravelly stones that then break down under different conditions into gravel soils. These soil types are very common in the south west.

Hot dry and hugely windy climates have seen comprehensive periods of erosion occur, where sands have been exposed and blown away to be deposited in deep systems as sandplains. These landform units now support one of our most diverse and famous floras known as the Kwongan, a Nyoongar Aboriginal term that literally means the rich heath-like vegetation of the sandplains in the south west. These few examples briefly describe but two parts of a very complex system of soil types that help make the many distribution patterns in this area's amazing flora possible.

Flowers in full bloom in the month of September, Gooseberry Hill, Perth.

4. An island situation, water to the west and south and desert to the north and east.

Together with being a very old land surface the south west is also an island in the sense that it has oceans to the south and west and deserts to the north and east. The oceans have been in existence for at least 65 million years when Australia separated from Antartica. The desert barriers have been becoming ever drier since around 25 million years ago until the establishment of real aridity around 5 million years ago, when dry climates and an increase in fire brought around the demise of the rainforest that had dominated since around 80 million years ago. Further, within the larger island phenomenon, the distribution of the flora as vegetation systems is also specific to soil type islands, the boundaries of which are also marked by sharply distinct variation in vegetation types. For example where sandplain meets creekline, low Kwongan heathland will all of a sudden give way to trees with a thicket understorey because the soil type has changed from sand over clay to deep sandy loam. This evidence tells us that the flora has evolved in isolation in a situation where plants have been forced by a range of conditions to become different species, subspecies and varieties, in an almost unparalleled manner.

5. Comprehensive long term availability of pollinating animals.

While this book helps introduce the concept of an incredibly rich flora, the associated animal life, principally insects, is even more diverse. Plants and animals have evolved together so that another of the factors that has driven the enormous diversity in the flora has been the actions and the opportunities brought about by the activities of pollinators. A huge percentage of the plants rely on pollinators to enact the process of fertilisation to ensure genetic strength and improved chances of survival. A study of the activities of nectar feeding animals on flowers will reveal features such as red or yellow, longer flowers with more nectar, attract honeyeating birds, while white shorter flowers with less nectar will attract insects. Some of the most fascinating and bizarre associations and adaptions have been discovered among observations of animals feeding on flowers. The more famous examples include:

Male Thynid wasps pollinating Drakea orchids.
The orchid flower not only looks like, but also smells like, a female wasp and the action of the male attempting to carry the flower away for copulation results in a flexible elbow on the flower shoving pollen onto the male wasp, who then visits another flower and cross pollination occurs.

Blowflies pollinating Hakeas.
Along the south coast a species of hakea, **Hakea rubiflora** or Stinking Roger, has brown and yellow flowers in spring that smell like rotting meat. When the bushes are in full flower they are laden with blowflies seeking an egg laying site and are cross pollinated by the flies visiting several plants.

PROTEACEAE - BANKSIA Etc.

Banksia media
(Southern Plains Banksia)

Grevillea decipiens

MYRTACAE - MYRTLES

Eucalyptus erythrocorys
(Illyarrie)

Calothamnus sanguinius

Dryandra formosa
(Showy Dryandra)

Lambertia propinqua

Melaleuca hamulosa

Calytrix decandra
(Pink Starflower)

Hakea laurina
(Pincushion Hakea)

Conospermum distichum

Kunzea ericifolia
(Spearwood)

Verticordia chrysantha

Isopogon latifolius

Adenanthos meisneri

Beaufortia schaueri
(Pink Bottlebrush)

Darwinia neildiana
(Fringed Bell)

FABACEAE
(PEA FAMILY)

Gastrolobuim bilobum
(Heart leaved poison)

Jacksonia hakioidies

Mirbelia dilatata
(Prickly Mirbelia)

Daviesia flexuosa

ORCHIDACEAE
(ORCHIDS)

Thelymitra variegata
Queen of Sheba Orchid

HAEMODORACEAE
(KANGAROO PAWS)

Anigozanthos rufus
(Red Kangaroo Paw)

CHLOANTHACEAE
(LAMBSTAILS)

Lachnostachys eriobotrya
(Lambswool)

EPACRIDACEAE

Andersonia grandiflora

MIMOSACEAE
(ACACIA)

Acacia heterochroa subsp.
heterochroa

GOODENIACEAE
(GOODENIAS)

Goodenia dyeri

Dampiera eriocephala
(Woolly headed Dampiera)

Scaevola crassifolia

ASTERACEAE
(DAISIES)

Schoenia cassiniana

LILIACEAE
(LILY)

Thysanotus patersoni
(Twining Fringe lily)

RUTACEAE

Boronia coriacea

STERCULIACEAE

Keraudrenia integrifolia
(Firebush)

19

Plant Names

With the benefits and pleasures derived from having a diverse flora here in the south west comes the complex problem of plant idenfication. The initial task may appear daunting but with a little perseverance, one can recognise the different characteristics between one species and another. This book has been designed to take the reader through the first and most basic stage of plant identification, that of simply comparing what you see in the field with the species illustrated in a book. It is not the best scientific technique but is the best way to familiarise yourself initially with the family groups we have. The first step may just be identifying the very common species highlighted in chapter 3 and marked with a red dot. But later you may notice that what you thought say was a simple **bottlebrush** could in fact be a **Kunzea** or **Calothamnus** or even a **Melaleuca** and if it is a melaleuca, is it *Melaleuca laterita* or *Maleleuca elliptica* or *Melaleuca calothamnoides*? All have superficial similarities to what we know as a **bottlebrush**. It is then that you realise the importance of knowing the Latin names of the various plants described.

Initially, the use of common names will assist in recognising certain plants, as this is a familiar vocabulary and it also allows the beginner to remember plant names by association of ideas; For example *Kennedia prostrata* is called **Running Postman**. So one can see that the plant in the field is red in colour and has a prostrate, creeping nature, but then another plant called *Gompholobium polymorphum* may be found which looks similar in shape and colour and has a similar prostrate nature but it has no common name. Furthermore it has a totally different scientific name, but one should not despair, as this is where the real joy of wildflower identification begins. That of simply starting to recognise the difference between one species and another, then success in identification is most rewarding. The importance of using Latin names will become evident the more one's interest develops. There are other advanced methods of plant identification such as 'keying out' species but this requires further study and is beyond the scope of this book. For further information on plant idenfication refer to the recommended reading list at the back of this book.

Every new plant identified by science is given a Latin name, first by genus and then by specie. The diagram below shows the basic structure of how *Eucalyptus marginata* subsp. *thalassica* (**Blue-leaved Jarrah**) is classified and named.

Angiosperms
(Flowering plants possessing naked seeds)

Dicotyledons (dicots)
(Having 2 seed leaves with floral
parts mostly in 5 sections, sometimes 4 sections.

Monocotyledons (monocots)
(Having 1 seed with floral
parts mostly in 3 sections)

Order (Myrtales)

Family (Myrtaceae)
(Contain genera that possess the same fundamental characteristic, including leaf oil glands in this family)

Genus (Eucalyptus)
(Group of species that share similar characteristics)

Species (Eucalyptus marginata)
(A plant that possesses certain specific characteristics)

Subspecies (thalassica)
(A plant differs in a basic characteristic feature but not substantially enough to warrant being classed as a separate species; in this case the plant differs from Eucalyptus marginata subspecies marginata by having uniformly bluish leaves)

The Structure of Flowers

Over millions of years flowers have evolved into varying complex shapes and structures. There is, however, a basic common structure that the majority of flowers possess and no matter how varied the shape, one can generally identify these similarities.

The diagram opposite illustrates a simplistic flower structure and identifies the main components within a flower. Adjacent to it are 4 examples showing how flowers can appear quite different but retain the same basic elements.

A flower normally has four main parts that are positioned on the receptacle and pedicel. The Perianth does not contain the reproduction components but primarily is designed to attract pollinators by the shape and colour of its combined petals and sepals.

The androecium made up of filaments and anthers form the male reproductive units. The pollen grains (male) are located in the anther head at the top of the stamen.

The gynoecium is made up of carpels having an ovary, style and stigma. Inside the ovary are ovules that contain the female egg cells. The stigma is designed to capture compatible male pollen and this is transferred to the ovules and fertilisation takes place. The ovule then transforms into a seed and the carpel develops into a fruit.

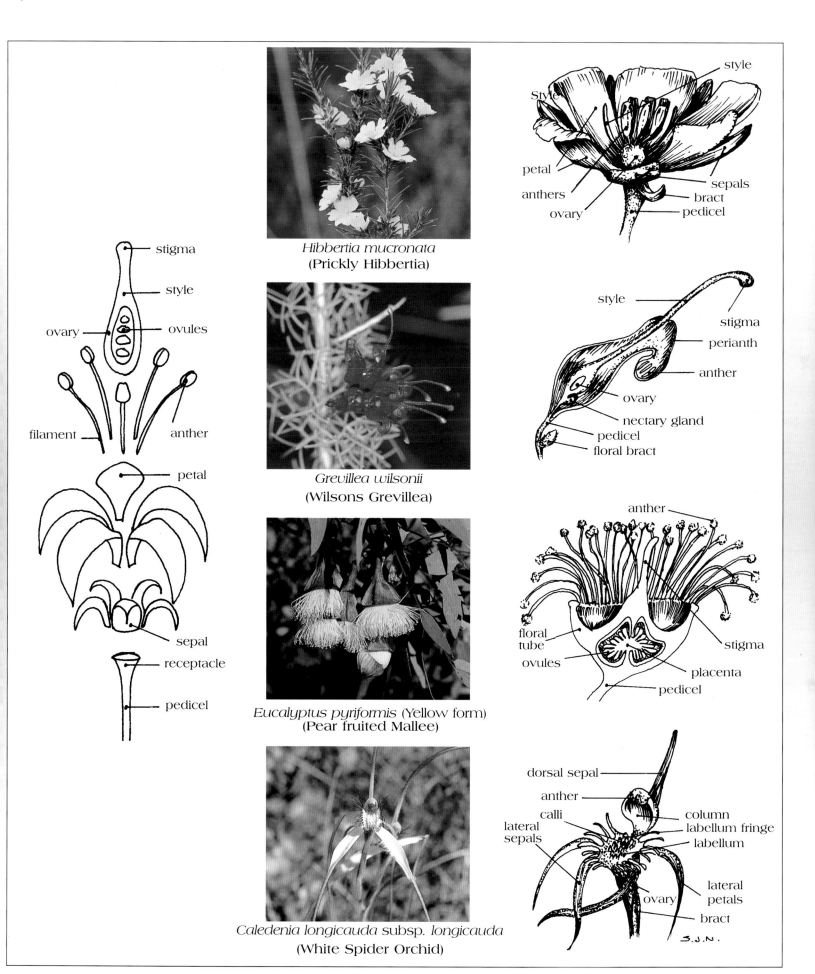

stigma

style

ovary

ovules

filament

anther

petal

sepal

receptacle

pedicel

Hibbertia mucronata
(Prickly Hibbertia)

style

Style

petal

anthers

ovary

sepals

bract

pedicel

Grevillea wilsonii
(Wilsons Grevillea)

style

stigma

perianth

anther

ovary

nectary gland

pedicel

floral bract

Eucalyptus pyriformis (Yellow form)
(Pear fruited Mallee)

anther

floral tube

ovules

stigma

placenta

pedicel

Caledenia longicauda subsp. *longicauda*
(White Spider Orchid)

dorsal sepal

anther

calli

lateral sepals

column

labellum fringe

labellum

lateral petals

ovary

bract

S.J.N.

For many of us, native flowers are simply objects of beauty. However, flowers are much more than they appear to the human eye and nose. Pollination ecology studies the interaction between plants and their specific pollination method. When you think about it, plant and animal survival depends on the integrity of this important process. Fertile, genetically healthy seeds must be produced and germinated at a naturally determined rate and density.

Most of our native flowering plants need to be pollinated. This task has been carried out for eons by native insects by day and by night and by birds during the day and mammals by night.

Receptors in the eyes of flower-visiting insects allow them to see in the ultraviolet end of the spectrum. An apparently white flower without obvious odour appears to its insect pollinator as pale blue with dark maroon splotches that guide the pollinator to the nectar source. Special odour receptors also allow insects to appreciate the subtle aromas of some plants. The pollination ecologist must consider many factors when ascertaining the diverse methods used by blossoms for pollination. Important features include: flower geometry, length of stamen, structure of anther and receiving stigma, the colour, the shape, and pattern of petals, sepals and bracts, and the arrangement of the individual blossoms on the plant. The timing of flowering, pollen presentation, receptivity of the female organ and order of blossoms opening are features that combine to form a complex and dynamic system that extends far beyond a human's admiring glance.

Blossoms may be grouped together on the basis of the type or 'syndrome' of their pollination. Methods of pollen transfer range from wind and water to mammals, birds and insects.

SOUTH WESTERN SYNDROMES

Pollination by water is mainly associated with marine and aquatic flowering plants. The genus Clematis of the family Ranunculaceae, is a rare example of a terrestrial rain-pollinated flowering plant.

Pollination by wind is mainly associated with the grasses, sedges and sedge-like plants (monocotyledons) and the dicotyledons sheoaks and chenopods (saltbushes, bluebushes and samphires).

Mammal-pollinated blossoms are typically large, strongly constructed, dull coloured, often near the ground and producing copious quantities of acrid smelling nectar at night. Typical examples include prostrate banksias and dryandras, some hakeas, grevilleas and eucalypts.

Bird-pollinated blossoms are typically large, strongly constructed, brightly coloured (reds, oranges, yellows and creams) and tubular. They may be positioned anywhere on the plant and produce copious quantities of sweet nectar during the day. Nectar guides are absent. Typical examples include bottlebrushes, one-sided bottlebrushes, some paperbarks, some eucalypts, bloodwoods, banksia, hakea and grevillea, native honeysuckles, woolleybushes, kangaroo paws, astrolomans and diploaenas.

Butterfly pollinated blossoms are uncommon in our semi-arid flora and reflect the scarcity of butterfly species. Moths, however, are predominantly nocturnal and are usually associated with small tubular flowers cream to white in colour, often clustered, nocturnally perfumed and generally richer in nectar than butterfly blossoms. Nectar guides are usually absent. Typical examples include Curry Flower, andersonia, boronia and banjines.

Beetle pollinated flowers are generally associated with large and diverse family Myrtaceae. Nectar guides are usually absent, blossoms are usually shallow and bowl-shaped and held erect with short, sturdy exposed sexual organs. Colour is often white or cream with rich non-sweet nectar. Favoured plant groups include eucalyptus, bloodwoods, paperbarks, ti-trees, scholtzia, baeckea and thryptomene. jewel beetles and cockchafers are conspicuous and attractive flower visitors.

Flies and midges visit a variety of blossom-types from simple bowl-shaped radial flowers to complex orchid blossoms. Odour ranges from sweet through to putrid. Nectar guides are sometimes present. Colour tolerance is also broad, though vivid reds are rarely visited. Specialist fly blossoms include briefly pollinated triggerplants, blowfly pollinated grevillea leucopteris and some hakeas, mosquito/gnat pollinated mosquito orchids, gnat-pollinated helmet orchids and midge-pollinated greenhood orchids.

Western Spinebill about to feed on Banksia sphaerocarpa.

Butterfly feeding on the flowers of a Grasstree (Xanthorrhoea pressii)

Bee gathering pollen off an Evergreen Kangaroo Paw (Anigozanthos flavidus)

Specialised wasp-pollinated blossoms are associated with some spider orchids, and all hammer orchids, slipper orchids, duck orchids, elbow orchids, and beard orchids.

These orchids use both aromatic and use visual deception to mimic the females of some groups of wasps. Nectar and nectar guides are absent.

There are numerous cases of native bee-pollinated blossoms in the South-West, ranging from peas and orchids to banksias and daisies. Some bees are specialised to a single species or genus. Nectar guides are present or absent. "Buzz pollination" is relatively common in the South-West solanum being a typical example. Some astrolomas and goodenias are bee specialists.

Western Australians are privileged to have such a wonderful diversity of plants and pollinators on their doorstep. However, we know little about their relationships. Moves are afoot to develop a plant-pollinator database which will be a valuable tool for all those interested in preserving native bushland.

Honeybees – The Unknown Quantity

Perhaps the greatest animal threat to our native pollinators is the Honeybee. It might come as a surprise to some that the Honeybee is introduced. In fact they were crucial to the success of white settlement in Australia becuase they pollinate most of our fruit and vegetables, as well as providing honey. It wasn't long before the first hive bees were allowed to form wild or feral populations. Providing three fundamentals were met: water all year round (permanent rivers, pools, lakes and dams), flowering most of the year (both native flora and weeds), and shelter in the form of tree hollows and rock crevices, feral honeybees exerted a new and formidable pressure on native pollinators.

Able to operate at lower temperatures than most native insect pollinators, bees forage for limited nectar and pollen resources up to two hours before and after native insects begin and end their daily activities. Research has indicated that when bees are present they compete with native bees (and probably many other native insects), which honeybees may be forced on other species. Exclude the honeybee and stressed natives will return to carry on their normal activities. Unfortunately in nature humans are not able to physically exclude feral bees and the pressure exerted over generations may push natives to the edge. Remember a beehive contains over 50,000 bees. Each hive divides every year when they swarm in spring. Both these hives then divide again next year, and so it goes until there are not enough hollows or crevices to shelter them. Of the many native creatures that use hollows none are able to repel a swarm, and many have young in spring. Death is a common outcome.

Small wheatbelt conservation reserves are particularly vulnerable when 10 hives (half a million bees) are brought in to graze on target flora. If there is a full complement of 10 feral hives already present then the natives pollinators must suffer the competition of one million foreigners where once there were none. This pressure may have been going on for generations in some areas causing local extinctions, before the full complement of native bee biodiversity has been sampled. If bees have displaced native pollinators one could expect a loss of the many species they do not pollinate properly, and a dependency by those they do pollinate, as honeybees are the only species left to do the job...!

There is something very satisfying about growing native plants. To witness the various stages in the development of a plant, from nurturing a seedling to seeing a mature plant producing flowers and seed, is one of the most rewarding pleasures.

When we choose to grow 'native' plants we are in fact assisting the environment by replacing those plants lost by land clearing. This allows fauna such as birds and insects to return to their original habitat and by replacing large areas of grassed lawns we help to conserve our precious water supplies.

Many species are incredibly easy to grow by germinating from seed and rearing under the most simple of conditions. Even those plants that were once difficult to grow are now made easier with the advent of 'smoke treatment', which triggers germination. Plants like the *Grevillea wilsonii* (illustrated on page 21) are now 'tamed' by this process.

Many shrubs that do not respond well to seeding methods may often strike easily from cuttings, such as some of our beautiful verticordias (commonly known as feather flowers).

For the beginner, using easy-to-germinate seeds is the way to begin. Then you can progress later by experimenting with cuttings or smoking techniques to increase your range of plants. Both the latter methods are easy enough but require more devotion and time.

A brief list of some of the easiest to grow genera are detailed below, but remember there may be the odd tricky species within each group.

Fabaceae (Pea Flowers)	*Aotus, Bossiaea, Chorizema, Daviesia, Gompholobium, Hardenbergia, Hovea, Jacksonia, Kennedia & Templetonia*
Myrtaceae (Bottlebrush Flowers and Waxes)	*Agonis, Beaufortia, Callistemon, Calothamus, Eremaea, Eucalyptus, Kunzea, Leptospermum Melaleuca and Regelia*
Mimosaceae (Wattles)	*Acacia*
Proteaceae	*Banksia, Dryandra, & Hakea*

Other genus types include: *Allocasuarina, Callitris, Diplolaena, Isolepis, Olearia, Rhagodia* & more including various monocots (ie Kangaroo Paw, lilly flowers, etc.).

With the exception of acacia and pea seeds, none of the above genera require any pre-treatment. Most acacia and peas are best treated by gently nicking the hard seed coat with secateurs to expose (but not damage) a tiny portion of the white germ inside. Alternatively, cover the seed with boiling water and soak for 30 mins prior to sowing.

Sowing may commence in autumn or spring when temperatures are mild. Best months are May-June or September-October. Spring sowing is safe however, as autumn germinated seedlings may be slow to establish or adversely affected during extremely harsh weather.

The main stages of growing plants are discussed below.

POTS, SOIL AND SOWING

Select pots or tubes that can be placed into a plastic holding tray approximately 35x30cm. Pots or tube sizes may vary from 5-7cm in diameter and from 7-14cm deep. You can fit from 20-40 pots into a holding tray depending on pot size. Deeper pots will provide a better root system for easier establishment once planted out. Alternatively, select trays with individual compartments; 48 pockets are ideal. The seedlings that grow within these segmented trays can be potted up and grown in larger containers or may be planted direct into the open. However, special care is needed during the first few months until a shallow root system establishes.

Pots must be clean. Old pots may be sterilised by washing and soaking in chlorinated water. Sodium hypochlorite (swimming pool liquid) is ideal if diluted in water. Soil must be clean and weed free. Plain, clean yellow sand will work but is heavy and dries out too quickly, so a good sterile native potting mix is the best choice. It is not necessary to buy 'seed raising mix'.

Fill pots with soil and 'tamp' down gently to create a level, slightly compacted surface, then gently water, taking care to retain a smooth surface. Fine seeds should then be sparingly sprinkled onto the surface and firmly pressed in. Large seeds should be pressed slightly below the surface. Gently water in. Subsequent watering is necessary to avoid the pot surface from drying out whilst seed is germinating. This may involve one or two waterings a day depending on conditions. Provided these procedures are followed, seedlings will emerge within 2-4 weeks.

To help avoid surface 'dry out' or heavy rain from washing out the seed, a very shallow layer of course river sand or fine gravel may be sprinkled onto the pot surface. Germinating seedlings can easily push through the gaps between the coarse particles.

PLACEMENT

Trays of sown pots may be placed in the open during mild weather, although it is safer to provide a sheltered enclosure covered in 50-70% shade cloth. Be sure the enclosure is in the open allowing full penetration of the sun and all available light.

Try to provide some protection from snails by raising the growing area above ground, or lay sawdust around the growing area. Snails will avoid having to cross this barrier to get to your plants. Snails can devour a precious tray of freshly germinated seedlings in one night and this is often the main cause of seedling loss.

GROWING ON AND PLANTING OUT

Many seedlings may emerge in each pot. Once they are over 1cm tall, remove excess seedlings leaving one strong specimen to mature. This method of growing is better than trying to transplant individual seedlings to new pots as it avoids unnecessary disturbance to growing seedlings.

Seedlings should thrive through the spring and summer period, but need constant watering. Once a day is sufficient, except in very hot conditions. Also beware of strong, dry winds which can cause rapid moisture loss. Plants will be ready for planting out in the autumn as soon as the first rains have soaked the soil surface. Place pots directly in the open one month prior to planting to harden them off. Planting is simply a case of inserting plants that have a good root system directly into the soil, but slightly lower so that there is a slight depression around the plant to hold excess water. Water in new plants generously.

There is no need to treat the soil or to add fertiliser. Most native plants prefer pristine or 'untampered' soils. Native plants may respond in some cases to feeding, but most will thrive if simply watered adequately. Many hardy species may survive and prosper without subsequent watering, but it is safer to water at least once a week during the first summer. No more watering will be necessary in the following years.

Some species fare better in alkaline soils, i.e. limestone and coastal sands, while others prefer the heavier soils of the South West forest or hills region. So, it is best to use species that are suited to your area. However, experiment with all your plants because many species adapt surprisingly well in conditions far removed from their native habitat. On a final note, you can take pleasure in knowing that you have contributed to helping restore some of our lost flora as well as hopefully creating a beautiful garden in the process.

1 · Shade- house enclosure.
Note: Adequate sun and light penetrating growing enclosure, plants above ground to provide protection from snails.

2 · Close up of growing trays.
Note: Various sizes of pots or segmented trays showing healthy seedlings

3 · Plants in open.
Note: *Mature seedlings hardening off in full sun.*

4 · A plant 'plug' from a segmented growing trays.
Note: *Illustrating 'plugs' with good root systems which may be planted direct or potted up.*

5 · Excess seedlings.
Note: *Excess seedlings are removed leaving one strong specimen to mature.*

6 · Simple equipment.
Note: *Shows holding trays, various pot sizes, segmented tray, slow release fertilizer for native plants, potting mix, secateurs, spatula.*

Fire

Australia has experienced fire for thousands of years and we know that Aborigines used fire to aid food gathering and they also knew that the new shoots from regenerating plants were a preferred food source for kangaroos.

The majority of our wild fires are started by lightning strikes but occasionally they are created by managed burns that get out of control, or by sheer vandalism.

Fire intensity, frequency and time of year affect plants in different ways. The Karri forest of the wetter regions can tolerate low intensity burning through the undergrowth but does not do well with high canopy fires. Jarrah on the other hand is not so damaged as Karri. Most of the species that grow in the Kwongan resprout from rootstock after fires, but some species such as the banksias and hakeas require a hot burn to release their seed and this may only occur after

several years. The habit of these plants to store the seed until fire or death of the plant but still flower annually is called 'bradyspory'. These plants have the benefit of distributing copious amounts of seed at an optimum period when additional nutrients are released to the soil assisting future germination. If the frequency of fires is too regular many plants will not have time to flower and produce seed. Also weeds have a greater chance of establishing themselves in areas too frequently burnt, as they are natuarally geared to disturbed soils. An exciting recent discovery is that smoke is a major stimulant for germination of many native plants. It can be applied in liquid form rather than through burning the bush.

There is still so much we do not know about fire management and it is a contentious issue that hopefully authorities will be able to find suitable solutions in the future.

Salinisation of soils

Salt has been present in soils for millions of years and is most visible in arid regions where the rainfall is insufficient to leach away the excess salts. This natural process is called 'primary salinisation' but there is also 'secondary salinisation' and this can be created by land clearing.

In the agricultural region of the south west, where 90% of the original vegetation has been cleared, there is some of the most extensive salinisation effected areas in Australia and the sad reality is that increased levels of salinity are greatly affecting those last areas of vegetation that remain.

Secondary salinisation has occurred for several reasons. As discussed previously much of the south west region is a flat ancient landscape with little or no major river systems, having basically a semi arid climate and a low rainfall. This has created the optimum conditions for secondary salinisation to occur and when the perennial native vegetation, which generally has deep root systems, is removed and replaced with shallow rooted annual crops, there is an increased rise in the water table. This is caused by a reduction of water absorption from the loss of native plants as the shallow rooted crops cannot reach the water table.

Damage occurs when excessive salts are leached to the surface due to there being no vegetation which would otherwise absorb water and salts. Over time the water table rises and leaches the salts to the surface soils and in turn is then mobilised by subsequent rains to the lower lying areas, affecting the root systems of trees and other vegetation, eventually killing them.

Luckily there has been much development in resolving the problem of secondary salinisation and Landcare groups, consisting of environmentally aware farmers and other like-minded people, have made great progress in replanting trees and creating irrigation channels to stop water run off. The Departement of Conservation and Land Management has also made major inroads into the problem of increased salinity and the preservation of Lake Toolibin east of Narrogin, is a classic example of cooperation with the Toolibin Community.

With present day remote sensing technology, it is sometimes possible to ascertain where the problem areas may occur in future years by the use of electromagnetic equipment in aircraft. From this information it is possible to map the areas affected and help towards the long- term planning of farmlands and reserves by either replanting trees or leaving that land uncleared.

Dieback

Dieback is basically a fungal disease that affects the cell structure within the roots and stems of a host plant. Over a period of time the fungus prevents the plant drawing up moisture and nutrients from the soil, eventually killing it. This may take quite a long time causing the plant to slowly lose leaves and hence the use of the term 'Dieback'.

There are several forms of Dieback but the most prevalent and destructive is *Phytophthora cinnamomi* and to a lesser degree other *Phytophthora sp.* Both affect areas in the deep south west, particularly where the annual rainfall exceeds 800mm, but even those areas with rainfall between 400mm and 800mm can be affected where summer rain and duplex soils occur.

Dieback affects a wide spectrum of plants but particularly the Proteaceae family which includes the banksias, grevilleas, hakeas and dryandras. some species are more susceptible than others, such as the rare *Banksia brownii* or Feather-leaved Banksia illustrated below.

The sad reality is that banksias for example are dominant plants within certain plant communities, being 'keystone species', which means that they play a major part in the ecosystem of an area and the survival of many birds and animals is dependent on them as a food source. When an area is greatly affected the consequences for the local fauna are quite devastating. There are, however, attempts to combat the disease and plants have been injected or sprayed with phosphoric acid, which has been successful in assisting certain plants to re-establish themselves.

If walking into areas particularly when wet check your boots and clean off excess mud, as the fungal spores are microscopic and you may have travelled from an infected area.

A weed - Cape Weed
Artotheca calendula

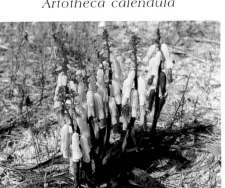

A weed -Soldiers
Lachenalia aloides

Feather leaved Banksia (*Banksia brownii*)
Killed by Dieback

A weed - Fountain Grass
Pennistelum setaceum

A weed - Purple Wood Sorrel
Oxalis purpurea

What is a weed ?

A weed is basically a plant that grows in an area where it is not wanted. In Western Australia they account for nearly 10% of the total flora.

The majority originate from overseas, particularly from South Africa, Europe and the Americas. Some of them are in fact Australian, having been introduced from other states.

Once established, some weeds will just remain in the area they were introduced to, but many, particularly the grasses, will spread far and wide as their seed is transferred by wind. Grasses are a particular problem as they invade and take hold in areas where the soil has been disturbed.

Why are weeds a problem? They contribute virtually nothing to the native flora and fauna ecosystem. Most native plants coexist in a complex, symbiotic relationship and weeds invade the soils between the native plants breaking this chain linkage, preventing native seeds germinating. Our precious roadside verges, that often contain rare flora, are being slowly enveloped by grasses and in some areas it is only a matter of time before they will totally dominate. Weeds produce little or no food for the native fauna, depriving them of their natural food source.

Farmers are well aware of the problems created by weeds, which cost the agricultural industry millions of dollars each year. They can poison stock, affect our waterways and reduce the environment to the most basic ecosystems.

We are all very aware of how time- consuming it is to remove weeds from a garden. Well, imagine the problems created in the bush. That is why we must maintain and protect our nature reserves, however small, as they retain the vast majority of our total native flora, and if soils are not disturbed, weeds find it hard to take hold and grow.

Eremophila virens

Snake Eremophila
Eremophila serpens

Grevillea tenuiloba

Grevillea prostrata

Mountain villarsia. *Villarsia calthifolia*

Our Roadside Verges

Besides our National Parks and Nature Reserves in Western Australia, one of our greatest assets is the extensive natural roadside verges that lie either side of some of our country roads. It was the Brand government in 1961 that had the foresight to set in place policies that required road verges to be at least 3 - 10 chains wide for the protection of flora and fauna, and when you drive up the Brand Highway or travel past the Lake Grace region, you will see these wide bands of natural vegetation between the cleared paddocks. Here many of our rare plants have been saved and of the 24 rare plants illustrated on these two pages, a few can only be found on road verges, highlighting the need to protect and preserve these important road reserves, which not only contain native flora but act as important corridors for the movement of birds and animals from one area to another.

The Roadside Conservation Committee has done much to protect these areas, as has the Wildflower Society of Western Australia, which has branches throughout the state and has done much to generate interest in wildflowers through their membership, creating such endeavours as 'The Bushland Conservation Fund', which finances revegetation programmes. So a dedicated few are doing much to protect what little remains.

The Wildflower Society of Western Australia is worth joining and there are sub groups that meet throughout the state. The head office address is - P.O. Box 64 Nedlands. W.A. 6909.

Grevillea scapigera

Siegfriedia darwinoides

Grevillea dryandroides subsp. *hirsuta.*

Banksia cuneata

Adenanthos labillardierei.

Grevillea concinna subsp. *concinna*

Grevillea involucrata

Lambertia orbifolia.
Round leaf Honeysuckle.

Adenanthos detmoldii.

Darwinea sp Mt Ney.

Acacia aphylla. Leafless Rock Wattle.

Daviesia oxylobium.

Road side verges completly cleared accept for one solitary Mottlecah.
(*Eucalyptus macrocarpa.)* Weeds will now proliferate.

Dryandra comosa.

Daviesia sp. Three Springs Daviesia.

A road verge left with the original native flora.

Daviesia euphorbioides.
Wongan Cactus.

Hakea aculeata. Column Hakea.

Eucalyptus rhodantha

Drummondita hasselli. var. *longifolia.*

Eucalyptus crucis.
Silver mallee.

The Vegetation Zones within the South West Botanical Province

The South West Botanical Province has been divided into seven vegetation zones in this guide book.

These zones are identified by their most obvious dominant vegetation types and follow the earlier work of Beard, Hopper and others.

The use of the dominant vegetation type for each zone also provides identifiable boundaries where the dominant type either becomes sparse and scattered, or disappears altogether.

The factors driving these patterns of vegetation distribution are changes in the underlying geologies, soil types and climate. The relationship between geology, soil and plants is known as an 'edaphic' relationship.

For example, Silver Princess, *Eucalyptus caesia* in its wild state only grows in the sandy loam soils found in the deeper pockets on some granite outcrops in the the Wandoo and Semi arid Woodland zones. Tuart *Eucalyptus gomophocephala* only grows on coastal limestone sands of the lower west coast in the banksia and eucalypt woodland zone.

The Seven broad vegetation zones outlined in this Chapter are as follows:

A further vegetation zone has been included, but lies outside the South West Botanical Province. It was included to assist those wishing to see the everlastings in the Mulga zone

Although these zones are named for the dominant vegetation types, several other vegetation types occur within their boundaries, including the "Kwongan" the botanically rich heath occuring on the sandplain in most zones.

Other smaller and botanically diverse features including granite outcrops, quartzite outcrops, salt lake systems and swamplands also occur through out the zones, all with their own unique flora.

How to use Chapter III.

When you are ready to plan a journey into the south west region, first refer to the map on page 5 to ascertain what vegetation zones you will pass through. After this you may refer to one of the detailed maps between pages 6 and 15 to locate specific areas you wish to visit. Then cross reference to the relevant vegetation zones in this chapter to see illustrated, some of the various splants you may encounter.

Under each photograph you will see the following layout:

genus: *Eucalyptus...* species: *torquata...* subspecies: (none)
common name: Coral Gum............. ●
botanic zone/s: Se .... J F M A M J J A S O N D
(common plant is identified by red dot)

The first line will have the scientific name in Latin starting with the genus and then species. Sometimes this will be followd by a subspecies name. The second line will have the common name. For many plants no common name exists. On the same line on the right hand side, you may see a red dot. This signifies when a plant is common either locally or throughout the various zones. This has been included to assist the beginner to identify the more common species. The third line starts with one or more of the following captial letters B, J, K, W, S, W, Se, M. This identifies the vegetation zone where the plant can be found. The letter references are shown in the opposite column e.g The Banksia Eucalypt Woodland is referred to as zone B.

On the same line on the right are the months of the year with the main flowering period highlighted in yellow. It must be remembered that this is a guide only, as many species may bloom out of season having intermittent flowering or in fact we do not know the full extent of the flowering period.

Within certain vegetation zones, you can find various sketch maps of flora rich reserves to assist you in your travels. Whenever possible do approach Conservation and Land Management or local Rangers to gain advice on where you can go. The flowers illustrated on the same page as the map will normally be located in that particular reserve but <u>not</u> <u>always.</u> They are, however, always in the vegetation zone selected.

As discussed previously, remember the picking of wildflowers without a licence is not allowed in this Western Australia - please leave them for others to enjoy.

Banksia Woodland showing Morrison Featherflower (*Verticordia nitens*) in full bloom in the first week of December and Slender Banksia (*Banksia attenuata*) Moore River National Park.

Banksia and Eucalypt Woodland

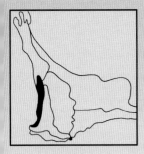

The Banksia and Eucalypt Woodland zone exists on the free draining soils to the west and north of the Jarrah Marri Forest zone.

The predominant landscape feature in this zone is the Swan Coastal Plain, a broad flat region running up the western edge of the Yilgarn Shield (granite) from the Whicher Range near Busselton in the south to beyond Gin Gin in the north, to where the Northern Mallee Heath zone on more complex soils takes over. It is from around ten to thirty kilometres wide between the Indian Ocean and the Darling Scarp, with limestone the principal underlying rock.

This zone is influenced by a relatively high rainfall zone from around 1000mm per annum in the south dropping to about 750mm per annum in the north.

Soil types within the zone can be categorised as predominantly sands. However, they fit into identifiable units of ever increasing age from the western-most recently formed Quindalup Dune System, followed by the progressively older Spearwood Dune System, Bassendean System, and through to the Pinjarra Plain, a colluvial (soils deposited downwards off slopes by gravity) sandy loam system situated up against the Darling Scarp (the western edge of the Yilgarn granite shield).

The common limestone outcropping over these western sand systems is known as the Cottesloe Limestone System, supporting most of the rarer and restricted plants of the zone.

This zone is one of the most affected landscapes in Western Australia, having suffered the clearing and fragmentation brought about by agriculture and urbanisation. However, some pockets of significant natural lands still occur, having enormous conservation value to the community.

The significant rainfall enables several vegetation types to exist on the sandy soils. The types are outlined as follows:

- the small although substantial **Tuart** (*Eucalyptus gomphocephala*) forest in the extreme south near Busselton

- mixed woodlands of banksia (**B.** *menziesii*, **B.** *attenuata*, **B.** *grandis*, and **B.** *prionotes*), **Jarrah, Tuart** and **Marri** stretching north.

- lowlands dominated by **Flooded Gum** (**E.** *rudis*), banksia (**B.** *ilicifolia*, **B.** *littoralis*) and paperbark (*Melalueca rhaphiophylla*)

- wetlands dominated by **Flooded Gums**, rushes (*Typha orientalis*), sedges (*Ghania sp.* and *Juncus sp.*) and several aquatic plants

- heathlands and shrublands overlying rocky areas of limestone, supporting luxuriant stands of parrotbush (*Dryandra sessilis*), **Honey myrtle** (*Melalueca huegelii*), an array of spectacular spring wildflower heaths and some important mallees such as **Fremantle Mallee** (*Eucalyptus foecunda*), **Limestone Mallee** (*E. petrensis*) and the rare and restricted **Yanchep Mallee** (*E. argutifolia*)

- some uncommon although significant vegetation types also occur within the zone; at its south eastern edge against the Whicher Range on low ironstone soils, heath and scrubland communities occur with very restricted endemic plants such as the **Ironstone Honey pot** *Dryandra nivea* subspecies and **McCutcheons grevillea** (*Grevillea mcutcheonii*)

- Against the Darling Range in the east, lies the narrow band of woodland community dominated by Wandoo and a beautiful small and crooked tree (*Eucalyptus lanepoolei*) occurs on the richer colluvial gravelly sandy loams brought down from the adjacent range over the millennia.

Banksia and Eucalypt Woodland

Banksia menziesii.
Menzies Banksia
J.B.N.　　J F M A M J J A S O N D

Banksia menziesii　Yellow form
Menzies Banksia
J.B.N.　　J F M A M J J A S O N D

Banksia ilicifolia
Holly - leaved Banksia.
J.B.　　J F M A M J J A O N D

Banksia littoralis.
Swamp Banksia.
J.B.N.S.　　J F M A M J J A S O N D

Banksia attenuata.
Slender Banksia.
J.B.N.W.S.　　J F M A M J J A S O N D

Nuytsia floribunda.
Christmas Tree.
J.B.N.W.S.Se.　　J F M A M J J A S O N D

Banksia laricina.
Rose Banksia.
B.　　J F M A M J J A S O N D

Banksia laricina fruit
Rose Banksia.
B.　　J F M A M J J A S O N D

Nuytsia floribunda
Christmas Tree
J.B.N.W.S.Se.　　J F M A M J J A S O N D

Banksia and Eucalypt Woodland

Grevillea thelemanniana

\B. [J|F|M|A|M|J|J|A|S|O|N|D]

Grevillea preissii. subsp. *pressii.*

B. [J|F|M|A|M|J|J|A|S|O|N|D]

Eremaea brevifolia.

B.N. [J|F|M|A|M|J|J|A|S|O|N|D]

Grevillea olivacea.

B. [J|F|M|A|M|J|J|A|S|O|N|D]

Melaleuca rhaphiophylla.
Swamp Paperbark.
J.B.N.W.S.Se. [J|F|M|A|M|J|J|A|S|O|N|D]

Anigozanthos manglessii.
Red and Green Kangaroo Paw.
N.B.J.W. [J|F|M|A|M|J|J|A|S|O|N|D] ●

Adenanthos meisneri.

B. [J|F|M|A|M|J|J|A|S|O|N|D]

Verticordia nitens.
Morrison Featherflower.
B. [J|F|M|A|M|J|J|A|S|O|N|D] ●

Hakea costata.
Ribbed Hakea.
B.N. [J|F|M|A|M|J|J|A|S|O|N|D] ●

Petrophile macrostachya.

B.N. [J|F|M|A|M|J|J|A|S|O|N|D]

Anigozanthos viridis.
Green Kangaroo Paw.
B. [J|F|M|A|M|J|J|A|S|O|N|D]

Verticordia densiflora.
Dense Featherflower. ●
B. [J|F|M|A|M|J|J|A|S|O|N|D]

Blancoa canescens.
Winter Bell.
B.N. [J|F|M|A|M|J|J|A|S|O|N|D]

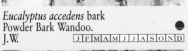

Eucalyptus accedens bark
Powder Bark Wandoo.
J.W. J F M A M J J A S O N D

Eucalyptus accedens.
Powder Bark Wandoo.
J.W. J F M A M J J A S O N D

Eucalyptus petrensis.
Limestone mallee.
B. J F M A M J J A S O N D

Eucalyptus marginata bark.
Jarrah.
B.J.S.W. J F M A M J J A S O N D

Corymbia haematoxylon.
Mountain Marri.
J. J F M A M J J A S O N D

Eucalyptus wandoo bark.
Wandoo.
B.J.W. J F M A M J J A S O N D

Corymbia calophylla bark.
Marri.
B.J.W. J F M A M J J A S O N D

Corymbia calophylla.
Marri.
B.J.W. J F M A M J J A S O N D

Eucalyptus rudis.
Flooded Gum.
B.J.S.W. J F M A M J J A S O N D

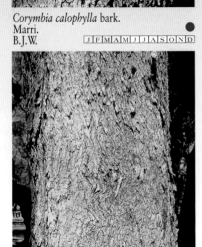

Eucalyptus gomphocephala bark.
Tuart.
B. J F M A M J J A S O N D

Eucalyptus gomphocephala.
Tuart.
B. J F M A M J J A S O N D

Eucalyptus decipiens.
Limestone Marlock.
B.S. J F M A M J J A S O N D

Eucalyptus decipiens.
Limestone Marlock.
B.S. J F M A M J J A S O N D

Some eucalypts of the Perth Region

Banksia and Eucalypt Woodland

Spinifex hirsutus
Hairy Spinifex.
B.N.S. JFMAMJJASOND

Spinifex longifolius.
Beach Spinifex.
B.N. JFMAMJJASOND

Hibbertia racemosa
Stalked Guinea Flower.
B. JFMAMJJASOND

Hibbertia pachyrrhiza.
B.J. JFMAMJJASOND

Thelymitra campanulata.
Shirt Orchid.
B.N.S. JFMAMJJASOND

Hovea stricta
B.N. JFMAMJJASOND

Oxylobium capitatum.
Bacon and Eggs.
B.N. JFMAMJJASOND

Cyrtostylis huegelii
Midge Orchid.
B.J.N.S.Se.W. JFMAMJJASOND

Halosarcia doleiformis.
Samphire.
B.W.S.Se.N.M. JFMAMJJASOND ●

Scholtzia involucrata.
Spiked Scholtzia.
B.N.J.W. JFMAMJJASOND ●

Pimelea floribunda.
A Banjine.
B.N. JFMAMJJASOND ●

Mesomelaena tetragona.
Semaphore Sedge.
N.B.J.S. · J F M A M J J A S O N D

Sowerbaea laxiflora
Purple Tassels
N.B.J. · J F M A M J J A S O N D

Hemiandra pungens.
Snakebush.
N.B.J.W. · J F M A M J J A S O N D

Thysanotus multiflorus
Many- flowered Fringe Lily
B.J.S. · J F M A M J J A S O N D

Eremaea fimbriata
N.B. · J F M A M J J A S O N D

Pelargonium littorale
B.J.S. · J F M A M J J A S O N D

Stirlingia latifolia
Blueboy
N.B.S. · J F M A M J J A S O N D

Hypocalymma angustifolium
White Myrtle
B.J.S.W. · J F M A M J J A S O N D

Calytrix lechenaultii
Starflower
B.N. · J F M A M J J A S O N D

Hardenbergia comptoniana.
Wild Wisteria.
B.J. · J F M A M J J A S O N D

Schoenoplectus validus.
Lake Club Rush.
B.S. · J F M A M J J A S O N D

Burchardia congesta
Milkweed.
B.J. · J F M A M J J A S O N D

Billardiera candida.
N.J.W.S. · J F M A M J J A S O N D

The Jarrah-Marri forest showing Jarrah *Eucalyptus marginata* in the foreground and Marri *Corymbia calophylla* on the right.

Jarrah - Marri Forest

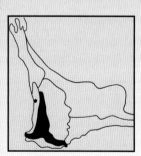

While Jarrah (*Eucalyptus marginata*) and Marri (*Corymbia calophylla*) both occur as component vegetation types within the Karri Tingle Tall Forest zone, these two trees occur as a dominant lower forest vegetation type in a zone of their own in the next highest rainfall region.

This relatively long linear zone adjoins the Karri Tingle Tall Forest zone in the south, running north and inland from and parallel to, the west coast on the south western edge of the Yilgarn shield. Here ancient monsoonal climates delivered laterite geological systems that have since weathered into relatively rich gravelly soils. subsequent climates have brought conditions creating a range of other soil types, which is now reflected by the modern vegetation types.

A typical contemporary Mediterranean climate sees the almost exclusively winter rainfall in this zone range from around 1300mm in isolated narrow strips in the western Darling Range east of Perth, down to around 850mm at its western edge and 700mm at its eastern and northern edges, where woodland zones begin.

The Jarrah Marri Forest zone, like all other zones, is made up of a mosaic of vegetation types. The others are:

Blackbutt (*Eucalyptus patens*), Flooded Gum (*E. rudis*) and River Banksia (*Banksia seminuda*) forest on lower damper loamy sites adjacent to drainage systems.

Wandoo (*E. wandoo*), Powderbark (*E. accedens*) and Buttergum (*E. laeliae*) woodlands on white and granite clays.

Low forest of paperbark (*Melaleuca rhaphiophylla*) and (*M. preissiana*), and Swamp Banksia (*B. littorea*) on winter-wet sandy sites.

Low woodlands and shrublands of banksia (*Banksia prionotes*), (*B. attenuata*), and Christmas tree (*Nuytsia floribunda*) on deeper sands.

Comprehensive and complex granite outcrop communities, rich shrublands, heathlands and herbfields on shallow soils and aquatic communities in rock pools.

Grevillea wilsonii.
Wilsons Grevillea.
J.B.   J F M A M J J A S O N D

Grevillea bipinnatifida.
Fuchsia Grevillea.
J.B.   J F M A M J J A S O N D

Calothamnus sanguineus.
Pindak.
B.J.S.N.W.   J F M A M J J A S O N D

Banksia grandis.
Bull Banksia.
J.B.W.S.   J F M A M J J A S O N D

Isopogon sphaerocephalus.
Drumstick Isopogon.
J.B.N.   J F M A M J J A S O N D

Melaleuca scabra.
Rough honeymyrtle.
J.B.N.W.S.Se.   J F M A M J J A S O N D

Lambertia multiflora subsp *darlingensis*
Many -flowered Honeysuckle.
J.   J F M A M J J A S O N D

Melaleuca radula.
Graceful Honeymyrtle.
J.B.N.W.Se.   J F M A M J J A S O N D

Adenanthos barbigerus
Hairy Jugflower.
J.B.   J F M A M J J A S O N D

Jarrah and Marri Eucalypt Woodland

Conostylis setosa
White Cottonheads.
B.J. ⬤ | J F M A M J J A S O N D |

Synaphea reticulata.
J. | J F M A M J J A S O N D |

Synaphea petiolaris.
J. | J F M A M J J A S O N D |

Pimelia suaveolens.
Scented Banjine.
W.J.S. ⬤ | J F M A M J J A S O N D |

Ptilotus manglesii.
Pom Poms.
J.W.Se.N. ⬤ | J F M A M J J A S O N D |

Persoonia saccata.
Pouched Persoonia.
B.J. | J F M A M J J A S O N D |

Persoonia microcarpa.
J.W.S.Se. | J F M A M J J A S O N D |

Conostylis breviscapa.
B.J.W. ⬤ | J F M A M J J A S O N D |

Comesperma confertum.
Milkwort.
B.N.S. ⬤ | J F M A M J J A S O N D |

Patersonia occidentalis.
Purple Flags.
J.B.W.S. ⬤ | J F M A M J J A S O N D |

Hibbertia subvaginata.
J.K. | J F M A M J J A S O N D |

Hibbertia hypericoides.
Yellow Buttercups.
J. ⬤ | J F M A M J J A S O N D |

Trymalium ledifolium var. *ledifolium.*
J. | J F M A M J J A S O N D |

Jarrah and Marri Eucalypt Woodland

Eucalyptus megacarpa.
Bullich.
J.B.

Xanthorrhoea gracilis.
Slender Grasstree.
J.B.W.S.

Kingia australis.
Drumsticks.
J.B.W.S.N.

Dasypogon hookeri.
Pineapple Bush.
K.J.

Persoonia longifolia
Long-leaved persoonia
J.

Allocasuarina humilis.
Dwarf Sheoak.
W.B.

Allocasuarina fraseriana
Sheoak
W.B.

Xanthorrhoea preissii
Grasstree
J.B.W.S.

Hakea amplexicaulis.
Prickly Hakea.
J.W.

Macrozamia riedlei
Zamia
J.B.S.

Hakea erinacea
Hedgehog Hakea
J.B.

Hakea undulata
Wavy-leaved Hakea
J.B.W.S.

Hakea trifurcata
Two-leaved Hakea
J.B.W.S.Se.N.

Jarrah and Marri Eucalypt Woodland

Cyanicula gemmata
Blue China Orchid
J.B.S. | J | F | M | A | M | J | J | A | S | O | N | D |

Caladenia footeiana
Crimson Spider Orchid
J.W.N. | J | F | M | A | M | J | J | A | S | O | N | D |

Caladenia filifera
Blood Spider Orchid
J.W. | J | F | M | A | M | J | J | A | S | O | N | D |

Calytrix fraseri
Pink Summer Calytrix
J. | J | F | M | A | M | J | J | A | S | O | N | D |

Acacia pulchella
Prickly Moses
J.B. | J | F | M | A | M | J | J | A | S | O | N | D |

Caladenia splendens
Splendid White Spider Orchid
B.W.S. | J | F | M | A | M | J | J | A | S | O | N | D |

Conospermum polycephalum
W.N. | J | F | M | A | M | J | J | A | S | O | N | D |

Conospermum huegelii
Slender Smokebush
J.W. | J | F | M | A | M | J | J | A | S | O | N | D |

Acacia alata
Winged Wattle
J.B.N.W. | J | F | M | A | M | J | J | A | S | O | N | D |

Glycine canescens
B.J.W. | J | F | M | A | M | J | J | A | S | O | N | D |

Acacia colletioides
Wait-a-While
N.Se.W. | J | F | M | A | M | J | J | A | S | O | N | D |

Acacia drummondii subsp. *affinis*
Drummonds Wattle
J.N.W. | J | F | M | A | M | J | J | A | S | O | N | D |

Acacia lateriticola
J. | J | F | M | A | M | J | J | A | S | O | N | D |

Jarrah and Marri Eucalypt Woodland

Thysanotus dichotomus
Branching Fringe Lily
J.B.W.N.S.Se. J F M A M J J A S O N D ●

Hybanthus floribundus
Showy Hybanthus
J.B.S.Se.N. J F M A M J J A S O N D

Hemigenia incana
Velvet Hemigenia
J.B. J F M A M J J A S O N D

Lechenaultia biloba
Blue Lechenaultia
J.B.W. J F M A M J J A S O N D ●

Thomasia glutinosa
Sticky Thomasia
J.S.Se. J F M A M J J A S O N D

Hypocalymma robustum
Swan River Myrtle
J.B.W. J F M A M J J A S O N D ●

Drosera stolonifera
B.J.W.N.S. J F M A M J J A S O N D ●

Utricularia menziesii
Red Bladderwort
J.B.W. J F M A M J J A S O N D

Lomandra odora
Tiered Mat Rush
J.B.S. J F M A M J J A S O N D

Johnsonia lupulina
Hooded Lily
J F M A M J J A S O N D ●

Actinotus humilis
Flannel Flower
J.B. J F M A M J J A S O N D

Drosera menziesii
Pink Rainbow
J.B.W.N.S.Se. J F M A M J J A S O N D ●

Stypandra glauca
Blind Grass
J.B.W.N.S.Se.M. J F M A M J J A S O N D ●

Jarrah and Marri Eucalypt Woodland

Andersonia lehmanniana

J. ⬛ | J F M A M J J A S O N D

Chamelaucium erythroflora.

J. | J F M A M J J A S O N D

Astroloma pallidum
Kick Bush

J.W.S. ⬛ | J F M A M J J A S O N D

Stylidium amoenum
Lovely Triggerplant

J. ⬛ | J F M A M J J A S O N D

Stylidium breviscapum
Boomerang Triggerplant

J.W. ⬛ | J F M A M J J A S O N D

Jacksonia restioides
Rush Jacksonia

J. ⬛ | J F M A M J J A S O N D

Scaevola calliptera

J.W. ⬛ | J F M A M J J A S O N D

Pericalymma ellipticum

J. | J F M A M J J A S O N D

Astroloma foliosum
Candle Cranberry

J.W. ⬛ | J F M A M J J A S O N D

Xylomelum occidentale
Forest Woody Pear

J. | J F M A M J J A S O N D

Podocarpus drouynianus
Wild Plum

J. ⬛ | J F M A M J J A S O N D

Baeckea camphorosmae
Camphor myrtle

J.W. | J F M A M J J A S O N D

Lechenaultia floribunda
Free Flowering Lechenaultia

B.J.W. | J F M A M J J A S O N D

Petrophile biloba
Granite Petrophile
J. | J F M A M J J A S O N D |

Petrophile serruriae
J.B.W.N.S.Se. | J F M A M J J A S O N D |

Isopogon asper
J. | J F M A M J J A S O N D |

Beaufortia purpurea
J. | J F M A M J J A S O N D |

Dryandra armata
Kangaroo Thorn
J.W.N.S. | J F M A M J J A S O N D |

Dryandra lindleyana
Couch Honeypot
J. | J F M A M J J A S O N D |

Dryandra sessilis
Parrot Bush
J.B.W.N.S.Se. | J F M A M J J A S O N D |

Pimelea lanata
J.B. | J F M A M J J A S O N D |

Grevillea manglesii subsp. manglesii
J. | J F M A M J J A S O N D |

Grevillea paniculata
J.W.Se. | J F M A M J J A S O N D |

Grevillea endlicheriana
Spindly Grevillea
J.W. | J F M A M J J A S O N D |

Darwinia citriodora
Lemon-scented Darwinia
J. | J F M A M J J A S O N D |

Verticordia acerosa
J. | J F M A M J J A S O N D |

Verticordia plumosa
Plummed Featherflower
J.W.S. | J F M A M J J A S O N D |

Conostylis setigera
Bristly Cottonheads
J.B.W.S.Se. | J F M A M J J A S O N D |

Anigozanthos flavidus
Evergreen Kangaroo Paw
J.K.S. | J F M A M J J A S O N D |

Jarrah and Marri Eucalypt Woodland

Daviesia decurrens
Prickly Bitterpea
J.W. | J | F | M | A | M | J | J | A | S | O | N | D |

Bossiaea eriocarpa
Common Brown Pea
W. | J | F | M | A | M | J | J | A | S | O | N | D |

Chorizema dicksonii
Yellow-eyed Flame Pea
J. | J | F | M | A | M | J | J | A | S | O | N | D |

Gastrolobium villosum
Crinkled-leafed Poison ●
J.W. | J | F | M | A | M | J | J | A | S | O | N | D |

Mirbelia dilitata
Holly-leaved Mirbelia ●
J.S. | J | F | M | A | M | J | J | A | S | O | N | D |

Gompholobrium polymorphum
J.W.S. | J | F | M | A | M | J | J | A | S | O | N | D |

Hovea chorizemifolia
Prickly Hovea ●
J. | J | F | M | A | M | J | J | A | S | O | N | D |

Hovea pungens
Devils Pins ●
J.W.N.B. | J | F | M | A | M | J | J | A | S | O | N | D |

Daviesa horrida
Prickly Bitterpea
J. | J | F | M | A | M | J | J | A | S | O | N | D |

Daviesia polyphylla
J.W. | J | F | M | A | M | J | J | A | S | O | N | D |

Nemcia spathulata
J.W. | J | F | M | A | M | J | J | A | S | O | N | D |

Gastrolobium spinosa ●
N.B.J.W.Se. | J | F | M | A | M | J | J | A | S | O | N | D |

Daviesia hakeoides
B.J.W. | J | F | M | A | M | J | J | A | S | O | N | D |

FABACEAE - Pea family

On the Wansbrough Walk, Porongurup National Park showing the tall Karri trees *Eucalyptus diversicolor*.

Karri and Tingle Forest

The extreme south west and adjacent south coast of Western Australia is prevailed upon by the wettest climate in the South West Botanical Province, where up to 1400 mm of rain falls annually. This combined with deeper soils off the southern edge of the Yilgarn shield makes this the region home of the largest and tallest forests in Western Australia.

Huge eucalypts dominate this vegetation zone, the smooth barked **Karri** *(Eucalyptus diversicolor)* the most widespread, occurring from near Cape Naturaliste to east of Albany and the Porongorup Range.

The three species of tingle occur as enormous, often buttressed rough barked trees, not as tall as Karri although much more massive. **Yellow Tingle** *(Eucalyptus giulfoylei)*, **Red Tingle** *(E jacksonii)* and the rarest **Rates Tingle** *(E brevistylis)*. Restricted to the very highest rainfall zone adjacent to the western south coast, they are components of a relictual forest, meaning that this is the last of a once much more widespread forest type that has shrunk as the climate has become drier and is now restricted to a few hundred square kilometres in the wettest and coolest part of the state.

The zone is also the home of a whole suite of associated vegetation types occurring as a complex mosaic. They include: Karri forest, Tingle forest, Jarrah forest, woodlands, swamplands, coastal heaths, granite outcrop and ironstone heath.

The Karri and tingle forest type contain several common endemic as well as more widespread understorey plants. **Marri** *(Corymbia calophylla)* grows in this vegetation type as a forest giant, as well as in the other woodland and coastal types within the zone as a smaller tree or even a windswept shrub on the coast.

The Jarrah forest is dominated by the largest Jarrah *(Eucalyptus marginata)* trees in the world.

Low Jarrah, banksia, Peppermint or Marri woodlands occur on the poorer free- draining sandy and quartzy soils, and **Yate** *(Eucalyptus cornata)* woodland on loamier sands around damper sites. **Peppermint** *(Agonis flexuosa)* woodlands often have very sparse understoreys of herbs and grasses while the other types have more luxuriant and complex understoreys. The famous and vibrant **Red Flowering Gum** *(Corymbia ficifolia)* occurs within these units often hybridising with **Marri** *(Corymbia calophylla)*.

The swamplands are particularly rich, making it the most diverse in the zone, dominated by the Myrtaceae family featuring some very spectacular plants, many flowering in summer and autumn.

Coastal heaths predominate on sands over limestone or granite, dominated by low windswept areas of shrubby plants.

Granite outcrops are scattered throughout the zone, frequently dominating the landscape as the highest topographical features, supporting their own unusual suite of plants, including fascinating aquatic flora in rock pools.

The peculiar ironstone heath overlies granular iron a little east of Cape Leeuwin in the south west of the zone. It is worthy of mention here as it shows how strongly the flora has adapted to the substrate differences, in this case iron rich soils. A plant of particular note here is the tall **Yellow Jug Flower** *(Adenanthos detmoldii)* of the damper low lying places, often on roadsides.

Grevillea depauperata prostrate form

J.K. | J F M A M J J A S O N D |

Kennedia coccinea
Coral Vine
K.J.B.S.N. | J F M A M J J A S O N D | ●

Utricularia multifida
Pink Petticoats
K.J.B.S.N. | J F M A M J J A S O N D |

Trymaliun spathulatum
Karri Hazel
K. | J F M A M J J A S O N D | ●

Allocasuarina decussata
Karri Sheoak
K. | J F M A M J J A S O N D | ●

Hovea elliptica
Tree Hovea
K.J.S. | J F M A M J J A S O N D | ●

Eucalyptus jacksonii
Red Tingle
K. | J F M A M J J A S O N D | ●

Acacia urophylla
K. | J F M A M J J A S O N D | ●

Scaevola sp.
K.J. | J F M A M J J A S O N D |

Pimelea brachyphylla
K.J. | J F M A M J J A S O N D | ●

Pterostylis barbata
Bird Orchid
J.K.S. | J F M A M J J A S O N D | ●

47

Karri and Tingle Forest

Kennedia prostrata
Running Postman
K.J. ⬤ JFMAMJJASOND

Leucopogon verticillatus
Tassel Flower
K.J.S. ⬤ JFMAMJJASOND

Chorilaena quercifolia
Chorilaena
K.J. ⬤ JFMAMJJASOND

Isotoma hypocrateriformis
Woodbridge Poison
K.B.J.Se.N.S.W. JFMAMJJASOND

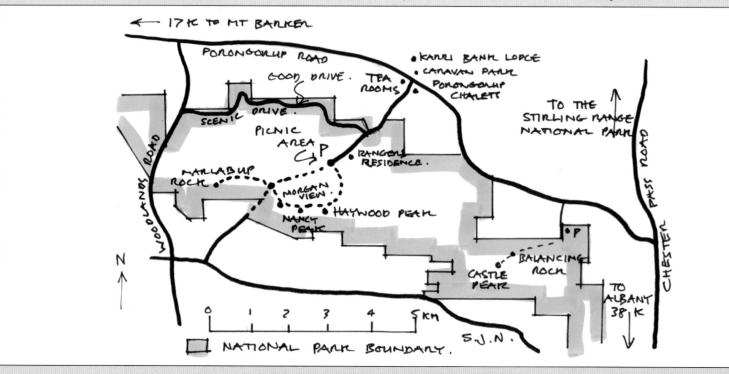

PORONGURUP NATIONAL PARK

Stylidium scandens
Climbing Triggerplant
J.K. ⬤ JFMAMJJASOND

Petrophile drummondii
K.J.B.W.Se. S.N. ⬤ JFMAMJJASOND

Hibbertia cuneiformis
Cut-leaf Hibbertia
K.J.B.S. ⬤ JFMAMJJASOND

Acacia applanata
K.J.B.W. JFMAMJJASOND

Open wandoo forest in Dryandra Reserve near Narrogin.

Wandoo Woodland

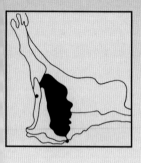

The Wandoo Woodland vegetation zone is named from its dominant and widespread woodland tree, the Wandoo (*Eucalyptus wandoo),* the name coming from the Aboriginal Nyoongar language.

While the range of wandoo extends outside this zone, mainly to the west on white clay (kaolin) soils, it is within that it occurs as the most widespread and dominant vegetation type.

This zone extends from the western and southern edge of the Jarrah Marri forest (where some of the tallest wandoo woodlands verge on being a forest because of the size of the trees), south-eastwards to the southern mallee shrublands and heath, north-eastwards to the somewhat similar semi-arid woodland and shrublands, and north to the northern mallee, shrublands and heath. The band is around 500km long and 150 km wide.

The climate is strongly Mediterranean with rainfall from around 650mm per annum in the west and south to around 450mm on the eastern and northern perimeter of the zone.

The underlying geology and soil types are all related to the existence of the granite Yilgarn Shield that dominates the zone. Soil types are deeper gravels in the west, interspersed with sand lenses and gravelly loams along drainage lines associated with the Darling Range. This grades to broad plains of alluvial loamy soils associated with ancient drainage systems, sand covered duplex soils, deep sand lenses, and contemporary saline drainage system tracts.

The zone has a very complex mosaic of associated vegetation types, including the western beginnings of some of the super rich Kwongan heathlands as extraordinary patches in the rich mosaic and the western limits of several vegetation types common in the next zones out. The types are:

- Wandoo Woodland

- Mallet eucalypt (*E. astringens* and *E. gardneri)* woodlands

- Salmon Gum (*E. salmonophloia*) York gum (*E. loxophleba)* and Morrel (*E. longicornis)* woodlands

- Marri jarrah woodlands

- Granite rock communities featuring the western populations of some of the most spectacular and restricted plants in the flora, Caesia (*E. caesia* subspecies *caesia* and yellow Sea Urchin Hakea (*Hakea petiolaris)*

- Sheoak (*Allocasuarina hugeliana*) and banksia (*B. prionotes*) and (*B. attenuata*) woodlands on deeper coarser sands, and Salt Water Sheoak (*Casuarina obesa*) along water courses

- Mallee shrublands and heath, featuring considerable plant diversity including many rare and restricted plants, and

- Kwongan, featuring the most spectacular and diverse vegetation type of the zone.

A magnificent anomaly sits between this and the next zone as the Wongan Hills and its surrounding plain, where a greenstone intrusion has uplifted through the Yilgarn Shield in antiquity and provided a mixture of deep gravels and outer yellow sandy clays. The mallee shrublands and Kwongan of this area are one of the most diverse hot spots in the whole South West area, with several endemic and very restricted plants present on the peculiar soils.

Wandoo Woodland

Petrophile squamata

J.B.W. | J F M A M J J A S O N D

Petrophile seminunda

J.W.S.Se. | J F M A M J J A S O N D

Petrophile rigida

W.S. | J F M A M J J A S O N D

Hakea lehmanniana
Blue Hakea

W.S. | J F M A M J J A S O N D ●

Leucopogon propinguus
Common Forest Heath

J.W.S. | J F M A M J J A S O N D

Diplolaena microcephala

W. | J F M A M J J A S O N D

Hakea petiolaris
Sea Urchin Hakea

J.W. | J F M A M J J A S O N D

Hakea ruscifolia
Candle Hakea ●

J.W.N.B.S. | J F M A M J J A S O N D

Dryandra polycephala
Many-headed Dryandra

J.W. | J F M A M J J A S O N D ●

Dryandra carduacea
Pingle

J.W. | J F M A M J J A S O N D ●

Dryandra squarrosa
Pingle ●

W.S. | J F M A M J J A S O N D

Dryandra ferruginea

W.Se. | J F M A M J J A S O N D ●

Dryandra cynaroides

W. | J F M A M J J A S O N D

Dryandra nobilis
Golden Dryandra

N.W. | J F M A M J J A S O N D

Banksia sphaerocarpa var. *sphaerocarpa*
Round Fruit Banksia ●

W.S.N. | J F M A M J J A S O N D

Beaufortia intestans

W.S. | J F M A M J J A S O N D

Wandoo Woodland

Tricoryne elatior
Autumn Lily
N.B.J.S.S.e.W. J F M A M J J A S O N D

Burchardia multiflora
Dwarf Burchardia
J.W. J F M A M J J A S O N D

Stackhousia huegelii
N.B.J.W.S.Se. J F M A M J J A S O N D

Cyanostegia corifolia
Tinsel Flower
W.Se. J F M A M J J A S O N D

Wurmbea tenella
Eight Nancy
B.W.J.Se. J F M A M J J A S O N D

Billardiera bicolor
Painted Billardiera
W. J F M A M J J A S O N D

Phebalium lepidotum subsp *tuberculosum.*
N.B.J.W. J F M A M J J A S O N D

Hibbertia subvaginata
J.W. J F M A M J J A S O N D

Borya constricta
Pincushion
W.J.Se.S. J F M A M J J A S O N D

Goodenia scapigera
White Goodenia
B.W.S.Se. J F M A M J J A S O N D

Scaevola lanceolata
N.W. J F M A M J J A S O N D

Thomasia macrocarpa
Large Fruited Thomasia.
W.S.Se. J F M A M J J A S O N D

Tripterococcus brunonis
Winged Stackhousia
J F M A M J J A S O N D

Drosera erythrorhiza
Red Ink Sundew
N.W.Se.S. J F M A M J J A S O N D

Boronia caerulescens
N.W.B.S.Se. J F M A M J J A S O N D

Casuarina obesa
Swamp Sheoak
W.N.Se.S. J F M A M J J A S O N D

Wandoo Woodland

Isopogon divergens
Spreading Coneflower
N.W.Se. ● J F M A M J J A S O N D

Diplolaena velutina
N.W. J F M A M J J A S O N D

Diplolaena microcephala
Lesser Diplolaena
N.W. J F M A M J J A S O N D

Lambertia ilicifolia
Holly-leaved Honeysuckle
W. J F M A M J J A S O N D

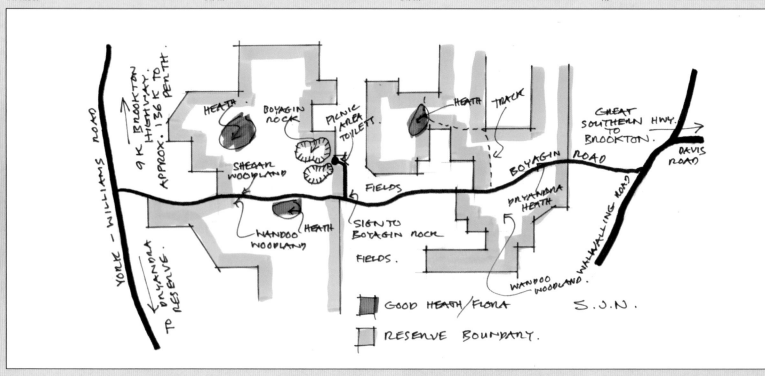

BOYAGIN NATURE RESERVE

Isopogon dubius
Rose Coneflower
B.W.S. ● J F M A M J J A S O N D

Isopogon formosus
Rose Coneflower
B.W.S. J F M A M J J A S O N D

Isopogon adenanthoides
Spider Coneflower
N.W. J F M A M J J A S O N D

Phebalium tuberculosum subsp.
tuberculosum
J F M A M J J A S O N D

Wandoo Woodland

Brachysema celsianum
Dark Pea Bush
W.S.Se. | J F M A M J J A S O N D

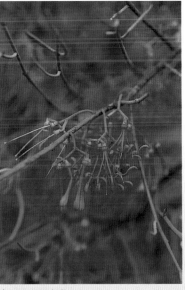

Amyema preissii
Narrow leaved Mistletoe
J.W. | J F M A M J J A S O N D ●

Billardiera erubescens
Red Billardiera
N.W.S. | J F M A M J J A S O N D

Adenanthos drummondii
N.B.W. | J F M A M J J A S O N D ●

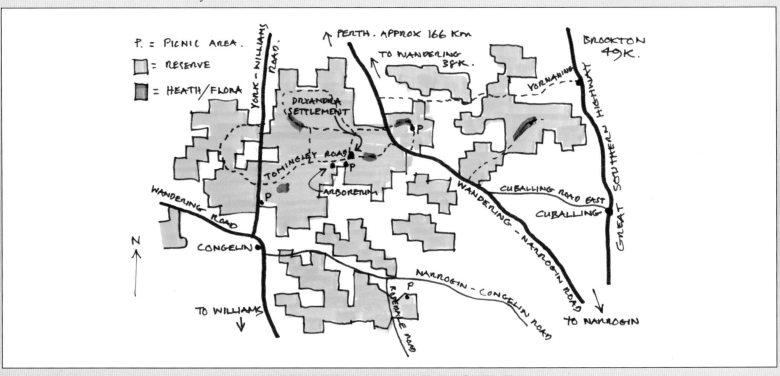

DRYANDRA NATURE RESERVE

Lechenaultia formosa
Red Lechenaultia
N.W.J.S. | J F M A M J J A S O N D ●

Pimelea ciliata
White Banjine
J.W. | J F M A M J J A S O N D ●

Astroloma pallidum. Dryandra form.
Kick Bush
J.W. | J F M A M J J A S O N D

Astroloma cilatum
Moss - leaved Heath
J.W. | J F M A M J J A S O N D ●

Wandoo Woodland

Lachnostachys ferruginea
Rusty Lambstails
W. JFMAMJJASOND

Regelia inops
W.S. JFMAMJJASOND

Anigozanthos humilis subsp *chrysanthus*
Mogumber Catspaw
W. JFMAMJJASOND

Anigozanthos humilis subsp. *grandis*
Giant Catspaw
W JFMAMJJASOND

Grevillea tenuiflora
Tassel Grevillea
W. JFMAMJJASOND

Grevillea leptobotrya. Dryandra form
Tangled Grevillea
W. JFMAMJJASOND

Grevillea kenneallyi
W JFMAMJJASOND

Hypocalymma puniceum
Large Myrtle
W.Se. JFMAMJJASOND

Allocasuarina huegeliana
Rock Sheoak
W.J. JFMAMJJASOND

Xanthorrhoea nana
Dwarf Grasstree
W.S. JFMAMJJASOND

Verticordia asterosa
J.W. JFMAMJJASOND

Verticordia lindleyi
W. JFMAMJJASOND

Eucalyptus caesia subsp *caesia*
W. JFMAMJJASOND

Eucalyptus celastroides subsp. *virella*
Mirret
W.Se. JFMAMJJASOND

Eucalyptus drummondi
Drummonds Gum
W.J. JFMAMJJASOND

Eucalyptus astringens Bark
Brown Mallet
W.S.Se. JFMAMJJASOND

Sphaerolobium sp.

J.W. `J F M A M J J A S O N D`

Hovea trisperma

B.J.S.W. `J F M A M J J A S O N D`

Daviesia gracilis

W. `J F M A M J J A S O N D`

Daviesia chapmanii
N.B.J.W.S.
Name `J F M A M J J A S O N D`

Daviesia incrassata

N.B.J.W.S. `J F M A M J J A S O N D` ●

Gompholobium knightianum
Handsome Wedge Pea
J.W. `J F M A M J J A S O N D`

Gastrolobium crassifolium

W. `J F M A M J J A S O N D`

Urodon dasyphylla
Mop Bushpea.
N.W.Se. `J F M A M J J A S O N D` ●

Templetonia biloba

W.J. `J F M A M J J A S O N D`

Daviesia cordifolia

W.S.Se. `J F M A M J J A S O N D`

Gastrolobium parvifolium
Berry Poison
J.W. `J F M A M J J A S O N D`

Gompholobium polymorphum
Variable Gompholoium
N.J.W.S. `J F M A M J J A S O N D` ●

Daviesia elongata subsp elongata

W.S. `J F M A M J J A S O N D`

Fabaceae - Pea family

Wandoo Woodland

Conospermum polycephalum
Common Smokebush
N.J.W.S.Se.B. ● J F M A M J J A S O N D

Pterostylis recurva
Jug Orchid
N.J.B.W.S.Se. ● J F M A M J J A S O N D

Caladenia falcata
Green Spider Orchid
W.S. J F M A M J J A S O N D

Diurus corymbosa
Common Donkey Orchid
N.B.J.W.S. ● J F M A M J J A S O N D

Prasophyllum macrotis
Inland Leek Orchid
W.Se.N. J F M A M J J A S O N D

Caladenia reptans subsp. *reptans*
Little Pink Fairy Orchid
N.B.J.W.S. J F M A M J J A S O N D

Drakonorchis barbarossa
Common Dragon Orchid
W.S. ● J F M A M J J A S O N D

Acacia sp.
W. J F M A M J J A S O N D

Acacia nervosa
Rib Wattle
B.J.W. J F M A M J J A S O N D

Acacia browniana var. *intermedia*
W. J F M A M J J A S O N D

Acacia lasiocarpa var. *sedifolia*
W.
Name J F M A M J J A S O N D

Acacia botrydion
W. J F M A M J J A S O N D

Acacia drummondi subsp. candolleana ●
K.J.W. J F M A M J J A S O N D

Acacia celastrifolia
Glowing Wattle
W. J F M A M J J A S O N D

Acacia squamata
W.J.
Name J F M A M J J A S O N D

Acacia chrysocephala
W.Place J F M A M J J A S O N D

Wandoo Woodland

Stylidium repens
Matted Triggerplant
W.J. J F M A M J J A S O N D

Synaphea flabellformis
J.W. J F M A M J J A S O N D

Conospermum brachyphyllum
W.N. J F M A M J J A S O N D

Xylomelum angustifolium
Woody Pear
N.W.Se. J F M A M J J A S O N D

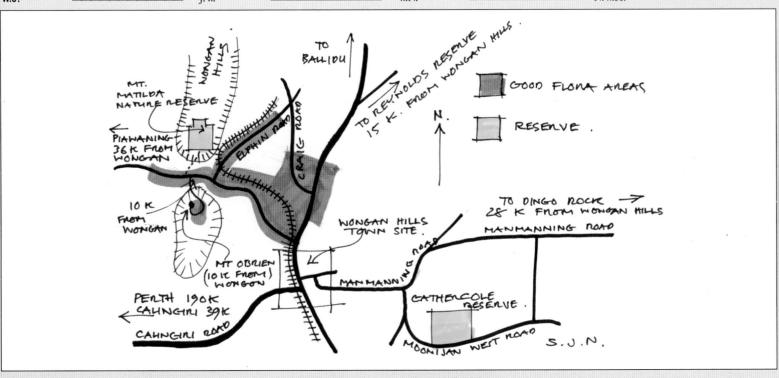

WONGAN HILLS REGION

Microcorys eremophiloides
W. J F M A M J J A S O N D

Beaufortia incana
W. J F M A M J J A S O N D

Conospermum ephedroides
W.Se. J F M A M J J A S O N D

Conostylis setigera
W. J F M A M J J A S O N D

Looking out to sea from Mermaid Point near Cheyne Beach east of Albany. In the foreground is the restricted Cutleaf Banksia (*Banksia praemorsa*).

Southern Mallee Shrubland and Heath

Botanical diversity across this extensive zone is enormous, influenced by a complex of variable and very large geological features.

Extending from the Jarrah Marri Forest zone in the south west and the Wandoo Woodland zone in the west, it narrows over 750 km eastwards to a coastal sliver out onto the Baxter cliffs on the Great Australian Bight To the north it runs into the Semi arid Eucalypt Woodland zone.

Rainfall is from around 750 mm in the extreme south west, to around 400 mm in the north west and drops off to 250 mm and less in the extreme east. The long southern coastal influence maintains rainfall up to 600 mm against the coast out to Cape Arid. The climate ranges from true Mediterranean in the west, to a less strongly dominated winter pattern towards the east where rainfall is less, although more likely to occur throughout the year.

The dominant geological features include:

- the southern edge of the Yilgarn shield
- Stirling Range and surrounding quartzite hills in the central west
- the Two Peoples Bay, Mt Manypeaks and Bremer Bay coastal granite systems
- the spongolite marine plain between the southern edge of the Yilgarn shield and the coast
- the Mt Barren quartzite ranges
- the greenstone intrusion into the Yilgarn shield of the Ravensthorpe Range
- coastal and offshore island granite systems of the Esperance area
- the eastern quartzite system of Mt Ragged and the Russell range, and
- towering coastal limestones of the western Nullabor cliffs.

This range of geology and landscapes provides a huge range of soil types, site aspects and local climates to support one of the greatest ranges of regional plant diversity in Australia.

Accordingly, the vegetation types represented in the zone vary comprehensively in diversity, in both composition and structure.

The vegetation types include:

- Diverse mallee shrublands with many dozens of mallee eucalypt species on a range of duplex soil types (duplex soils are those with a layer of usually sand over clay subsoils.)
- Kwongan heathlands massively diverse, on the deeper sandy and gravelly duplex soils. They vary in composition, following subtle changes in soil types, to the extent that sites within 5 km can have as little as 30% commonality in taxa present, even on very similar soils.
- Dense shrublands dominated by the Proteacae family, typically banksia, dryandra and lambertia. Massive nectar production makes these systems key food resource providers for animals. Again, species composition is altered by subtleties in soil makeup and large differences over short distances are commonplace.
- Coastal thickets predominantly of members of the Myrtacae family on stabilsed dune systems.
- Mallet eucalypt woodlands. Small trees that reproduce only from seed following fire, flood or storm type disturbance, often forming dense stands of one or a few species, and are very common on certain gravelly and loamy soils across the zone. The greatest diversity of mallet eucalypts occur in this zone with several dozen taxa present.
- Eucalypt woodlands of predominantly yate, (*Eucalyptus occidentalis*) along water courses and swamps, and the southern populations of Salmon Gum (*E salmonophloia*).
- Scattered hotspots of enormous endemism are commonplace, in particular the Stirling Range, the Fitzgerald River National Park area on quartzite and spongolite systems, the Ravensthorpe Range and the Esperance granites.

This zone offers the most dramatic examples of how complex and spectacular the combination of geological, landscape and soil variety can be in providing botanical diversity in the South West Botanical Province.

Southern Mallee Shrublands and Heath

Beaufortia orbifolia
Ravensthorpe Bottlebrush
S. | J | F | M | A | M | J | J | A | S | O | N | D |

Beaufortia sparsa
Swamp Bottlebrush
S.J.K. | J | F | M | A | M | J | J | A | S | O | N | D |

Beaufortia incana
J.S. | J | F | M | A | M | J | J | A | S | O | N | D |

Callistemon speciosa
Albany Bottlebrush
S. | J | F | M | A | M | J | J | A | S | O | N | D |

Callistemon phoeniceus
Lesser Bottlebrush
S.N.W. | J | F | M | A | M | J | J | A | S | O | N | D |

Kunzea baxteri
Baxters Kunzea
S. | J | F | M | A | M | J | J | A | S | O | N | D |

Calothamnus villosus.
Mouse ears
J.S.W.Se. | J | F | M | A | M | J | J | A | S | O | N | D |

Calothamnus gibbosus
Mouse ears
S. | J | F | M | A | M | J | J | A | S | O | N | D |

Beaufortia anisandra
S.Se. | J | F | M | A | M | J | J | A | S | O | N | D |

Southern Mallee Shrublands and Heath

Caladenia flaccida subsp. *pulchra*
Slender Spider Orchid
S.W. | J F M A M J J A S O N D

Caladenia polychroma
Common Spider Orchid
N.B.J.W.S. | J F M A M J J A S O N D

Caladenia longicauda subsp. *longicauda*
White Spider Orchid
B.J.S. | J F M A M J J A S O N D

Caladenia hirta subsp. *rosea*
Pink Candy Orchid
N.W.S.Se. | J F M A M J J A S O N D

Caladenia longiclavata
Clubbed Spider Orchid
B.J.W.S. | J F M A M J J A S O N D

Microtis media subsp. *media*
Common Mignonette
N.Se.W.B.J.S. | J F M A M J J A S O N D

Epiblema grandiflorum subsp *grandiflorum*
Babe in a cradle
| J F M A M J J A S O N D

Cyanicula sericea
Silky Blue Orchid
J.B.S. | J F M A M J J A S O N D

Pyrorchis nigrans
Red Beaks
N.B.J.Se.S. | J F M A M J J A S O N D

Microtis densiflora
Dense Mignonette
B.J.W.S. | J F M A M J J A S O N D

Prasophyllum brownii
Christmas Leek Orchid
J.W.S. | J F M A M J J A S O N D

Thelymitra villosa
Custard Orchid
J.S. | J F M A M J J A S O N D

Diuris pauciflora
Swamp Donkey
B.J.W.S. | J F M A M J J A S O N D

Diuris laxaiflora
Bee Orchid
N.B.J.S.Se.W.. | J F M A M J J A S O N D

Leptoceras menziesii
Rabbit Orchid
B.J.W.S.K.N.Se. | J F M A M J J A S O N D

Orchids

There are two main groups of orchids, epiphytes which mainly grow in trees and terrestrial orchids that grow in the soil.

Over half of the 900 orchids of Australia are of the epiphyte type, but in the south west, they are all terrestrial, there being over 300 species in this region.

The best time to find orchids is between August and November. They occur in varying habitats but around granite, outcrops, Sheoak woodland, Wandoo woodland and areas recently burnt are good places to look.

J F M A M J J A S O N D

ORCHIDACEAE - ORCHIDS

Adenanthos argyreus
Little Woollybush
S. ● J F M A M J J A S O N D

Adenanthos obovatus
Basket Flower
B.J.S. ● J F M A M J J A S O N D

Adenanthos flavidiflorus
S.Se.W. J F M A M J J A S O N D

Adenanthos cuneatus
J.S. J F M A M J J A S O N D

Adenanthos cygnorum subsp. *cygnorum*
S. J F M A M J J A S O N D

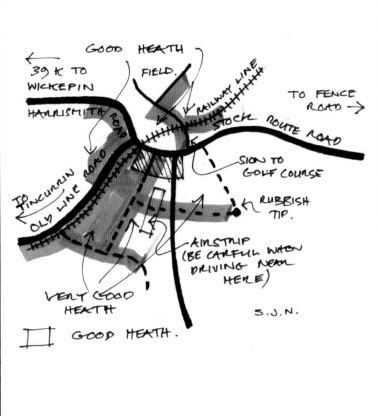

HARRISMITH REGION

Petophile longifolia
Long-leaved Cone Bush
S.W. ● J F M A M J J A S O N D

Adenanthos venosus
S. ● J F M A M J J A S O N D

Petrophile glauca
S. J F M A M J J A S O N D

Darwinia citriodora
Lemon-scented Darwinea
S.J. J F M A M J J A S O N D

Nematolepis phebalioides
S.Se.W. J F M A M J J A S O N D

Conostylis argentea
S.Se.W. J F M A M J J A S O N D

Petrophile heterophylla
Variable leaved Coneflower
S.Se.W. J F M A M J J A S O N D

Dryandra falcata
Prickly Dryandra
S. J F M A M J J A S O N D

Dryandra obtusa
Shining Honeypot ●
S. J F M A M J J A S O N D

Dryandra armata
Prickly Dryandra ●
N.W.J.S. J F M A M J J A S O N D

Dryandra quercifolia
Oak-leaved Dryandra ●
S. J F M A M J J A S O N D

Dryandra horrida
Prickly Dryandra ●
S.Se. J F M A M J J A S O N D

Dryandra cuneata
Wedge-leaved Dryandra ●
S.Se. J F M A M J J A S O N D

Dryandra foliosissima
Shaggy Dog Dryandra
S.Se. J F M A M J J A S O N D

Dryandra erythrocephala
S. J F M A M J J A S O N D

Dryandra baxteri
S. J F M A M J J A S O N D

Isopogon polycephalus
Clustered Coneflower ●
S. J F M A M J J A S O N D

Isopogon baxteri
Stirling Range Coneflower
S. J F M A M J J A S O N D

Isopogon scabriusculus subsp. *stenophyllus*
S J F M A M J J A S O N D

Isopogon teretifolius
S J F M A M J J A S O N D

Isopogon buxifolius var. *spathulatus*
S. J F M A M J J A S O N D

Isopogon axillaris
S.Se. J F M A M J J A S O N D

Isopogon trilobus
Barrel Coneflower
S.Se. J F M A M J J A S O N D

Hakea pandanicarpa
S.Se. ●
J F M A M J J A S O N D

Hakea nitida
Frog Hakea
N.B.J.W.S.
J F M A M J J A S O N D

Hakea corymbosa
Cauliflower Hakea
S.Se.W. ●
J F M A M J J A S O N D

Hakea cinerea
Ashy Hakea
S. ●
J F M A M J J A S O N D

Hakea clavata
Coastal Hakea
S. ●
J F M A M J J A S O N D

Hakea strumosa
S.Se.
J F M A M J J A S O N D

Hakea denticulata
Stinking Roger
S ●
J F M A M J J A S O N D

Hakea obtusa
S.
J F M A M J J A S O N D

Hakea sulcata
Furrowed Hakea
N.B.J.W.S.Se.
J F M A M J J A S O N D

Hakea ceratophylla
Horned Leaf Hakea
S.
J F M A M J J A S O N D

Hakea marginata
S.
J F M A M J J A S O N D

Hakea baxteri
S.
J F M A M J J A S O N D

Hakea varia
S.
J F M A M J J A S O N D

HAKEA

Southern Mallee Shrublands and Heath

Banksia solandri
Stirling Range Baksia
S. | J F M A M J J A S O N D |

Banksia aculeata
S. | J F M A M J J A S O N D |

Kunzea recurva
Mountain Kunzea
S. | J F M A M J J A S O N D |

Eucalyptus lehmannii
S.W. | J F M A M J J A S O N D |

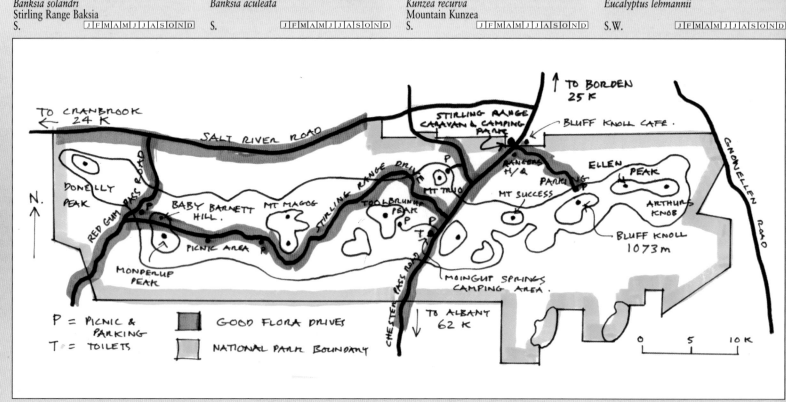

STIRLING RANGE NATIONAL PARK

Hibbertia stellaris
Orange Stars
S. | J F M A M J J A S O N D |

Lambertia ericifolia
Heath-leaved Honeysuckle
S. | J F M A M J J A S O N D | ●

Lambertia uniflora
S. | J F M A M J J A S O N D | ●

Hakea cucullata
S | J F M A M J J A S O N D | ●

Southern Mallee Shrublands and Heath

Isopogon cuneatus
A Coneflower
S. |J|F|M|A|M|J|J|A|S|O|N|D| ●

Beaufortia decussata
Gravel Bottlebrush
S.J. |J|F|M|A|M|J|J|A|S|O|N|D| ●

Beaufortia heterophylla
Stirling Range Bottlebrush
S. |J|F|M|A|M|J|J|A|S|O|N|D| ●

Darwinia meeboldii
Cranbrook Bell
S. |J|F|M|A|M|J|J|A|S|O|N|D|

Darwinia wittwerorum
Witters Mountain Bell
S. |J|F|M|A|M|J|J|A|S|O|N|D|

STIRLING RANGE
Looking east from Barnett Hill.

Gompholobium villosa
S. |J|F|M|A|M|J|J|A|S|O|N|D|

Andersonia echinocephala
Giant Andersonia
S. |J|F|M|A|M|J|J|A|S|O|N|D|

Hypocalymina speciosum
S. |J|F|M|A|M|J|J|A|S|O|N|D|

Darwinia oxylepis
Gillams Bell
S |J|F|M|A|M|J|J|A|S|O|N|D|

Acacia baxteri
Baxters Wattle
S. |J|F|M|A|M|J|J|A|S|O|N|D| ●

Orthrosanthus laxus
Morning Iris
N.J.W.S. |J|F|M|A|M|J|J|A|S|O|N|D|

Andersonia caerulea
J.S. |J|F|M|A|M|J|J|A|S|O|N|D| ●

Acacia gonophylla
S.   JFMAMJJASOND

Acacia delphina
S   JFMAMJJASOND

Acacia subcaerulea
S.   JFMAMJJASOND

Acacia durabilis
S.   JFMAMJJASOND

Acacia maxwellii
S.   JFMAMJJASOND

Acacia empeliolada
S.   JFMAMJJASOND

Acacia cephala
S.   JFMAMJJASOND

Acacia cephala
S.   JFMAMJJASOND

Acacia caerula
S.   JFMAMJJASOND

Acacia moirii subsp. *recurvistipula*
S.   JFMAMJJASOND

Acacia sulcata var. planifolia
S.   JFMAMJJASOND

Acacia moirii subsp. *moirii*
S.   JFMAMJJASOND

Acacia shuttleworthii
S
Name   JFMAMJJASOND

Acacia sulcata var. platyphylla
S.   JFMAMJJASOND

Acacia glaucoptera
Flat leaved Wattle   ●
S.W   JFMAMJJASOND

Acacia viscifolia
S.   JFMAMJJASOND

Kunzea micromera
S.

Beaufortia micrantha
Little Bottlebrush
B.S.Se.W.

Astroloma epacridis
S.

Pimelea rosea
Rose Banjine
B.J.S.

Alyogyne huegelii
Lilac Hibiscus
N.B.S.Se.W.

Isopogon polycephalus
Clustered Coneflower
S.

Johnsonia teretiflolia
Hooded Lily
S.

Agrostocrinum scabrum
False Blind Grass
B.J.S.W.Se.

Pimelea spectablis
Bunjong
S.J.

Pimelea crucens subsp. *crucens*
S.

Euphorbia paralias
Milkweed Spurge
S.

Dasypogon bromeliifolius
Drumsticks
B.J.S.

Pteridium esculentum
Bracken Fern
J.S.

Persoonia helix

Actinodium cunninghamii
Swamp Daisy
J.S.

Xanthosia rotundifolia
Southern Cross
J.S.

Corymbia ficifolia
Red- flowering Gum
S. J F M A M J J A S O N D

Eucalyptus tetraptera
Four- winged Mallee
S. J F M A M J J A S O N D

Eucalyptus nutans
Red -flowered Mort
S. J F M A M J J A S O N D

Eucalyptus pressiana
Bell Fruited Mallee
S. J F M A M J J A S O N D

Eucalyptus coronata
Crowned mallee
S. J F M A M J J A S O N D

Eucalyptus flocktoniae
Merrit
S.W.Se. J F M A M J J A S O N D

Eucalyptus desmondensis
S. J F M - M J J A S O N D

Eucalyptus burdettiana
Burdetts Mallee
S. J F M A M J J A S O N D

Eucalyptus macrandra
Long-flowered Marlock
S. J F M A M J J A S O N D

Eucalyptus megacornuta
Warted Yate
S. J F M A M J J A S O N D

Eucalyptus conferruminata
Bald Island Marlock
S. J F M A M J J A S O N D

Eucalyptus tetragona
Tallerak
N.W.S. J F M A M J J A S O N D

Eucalyptus sepulcralis
Weeping Gum
S. J F M A M J J A S O N D

EUCALYPTUS

Southern Mallee Shrublands and Heath

Agonis flexuosa
Peppermint
J.K.S. ● | J F M A M J J A S O N D |

Agonis marginata
Arnica
S | J F M A M J J A S O N D |

Agonis spathulata
S | J F M A M J J A S O N D |

Persoonia teretifolia
S. | J F M A M J J A S O N D |

Dampiera parvifolia
Many bracted Dampiera
S. | J F M A M J J A S O N D |

Melaleuca pulchella
Claw honeymyrtle
J.W.S. | J F M A M J J A S O N D |

Lysinema ciliatum
Curry Flower
N.B.J.S.W.Se. | J F M A M J J A S O N D |

Borya sphaerocephala
Pincushions
S.SeW. | J F M A M J J A S O N D |

Platysace compressa
Tapeworm plant
S.W.Se. | J F M A M J J A S O N D |

Lasiopetalum compactum
S. | J F M A M J J A S O N D |

Comesperma scoparium
Broom Milkwort
N.J.B.S.Se.W. | J F M A M J J A S O N D |

Chamelaucium virgatum
S.Se. | J F M A M J J A S O N D |

Sphenotoma dracophylloides
Paper Flower
S. | J F M A M J J A S O N D |

Pomaderris racemosa
Cluster Pomaderris
S. | J F M A M J J A S O N D |

Franklandia fucifolia
Lanoline Bush ●
J.S.Se.W. | J F M A M J J A S O N D |

Leucopogon sprengelioides
S. | J F M A M J J A S O N D |

Grevillea fastigiata

S. J F M A M J J A S O N D

Grevillea rigida. subsp. *distans*

S. J F M A M J J A S O N D

Grevillea baxteri. orange flowered form
Cape Arid Grevillea

S. J F M A M J J A S O N D

Grevillea excelsior
Flame Grevillea

S.Se.W. J F M A M J J A S O N D

Grevillea aneura

S. J F M A M J J A S O N D

Grevillea insignis subsp. *insignis.*
Wax Grevillea

S. J F M A M J J A S O N D

Grevillea macrostylis.
Mount Barren Grevillea

S. J F M A M J J A S O N D

Grevillea macrostylis small- leaved form.

S. J F M A M J J A S O N D

Grevillea tripartita

S. J F M A M J J A S O N D

Southern Mallee Shrublands and Heath

Grevillea fasciculata

S. ⛶JFMAMJJASOND⛶

Grevillea fistulosa

S. ⛶JFMAMJJASOND⛶

Grevillea nudiflora

S. ⛶JFMAMJJASOND⛶

Grevillea patentiloba

S. ⛶JFMAMJJASOND⛶

Grevillea acuaria

S. ⛶JFMAMJJASOND⛶

Grevillea asteriscosa
Star leaved Grevillea

S. ⛶JFMAMJJASOND⛶

Grevillea pectinata
Comb-leaved Grevillea

S. ⛶JFMAMJJASOND⛶

Grevillea dolichopoda

S. ⛶JFMAMJJASOND⛶

Grevillea superba

S. ⛶JFMAMJJASOND⛶

Templetonia sulcata
Centaipede Bush
S.

Chorizema glycinifolium
A Flame Pea
B.J.S.W.

Gompholobium capitatum
Yellow Pea
B.J.S.

Jacksonia compressa
S.

Urodon dasyphyllus
Mop Bushpea
S.W.Se.

Daviesia ovata
Broad leaf Daviesia
S

Mirbelia floribunda
Purple Mirbelia
Name

Daviesia rhombifolia
S.J.

Pultenaea verruculosa
S.

Gastrolobium velutinum
Stirling Range Poison
S.

Gompholobium venustum
Handsome Wedge Pea
B.J.S.W.Se.

Jacksonia spinosa
S.

Nemcia coriacea
S.

Eutaxia obavata
S.

Daviesia mollis
J.S.

Daviesia priessii
S.

FABACEAE - Pea family

Daviesia angulata

S.Se. | J F M A M J J A S O N D |

Gompholobium scabra
Painted Lady
B.J.S. | J F M A M J J A S O N D | ●

Jacksonia raccemosa

S. | J F M A M J J A S O N D |

Daviesia oppositifolia
Rattle Pea
S. | J F M A M J J A S O N D | ●

Templetonia retusa
Cockies Togues
N.B.J.S. | J F M A M J J A S O N D | ●

Nemcia leakiana
Mountain Pea
S. | J F M A M J J A S O N D | ●

Chorizema aciculare
Needle-leaved Flame Pea
S. | J F M A M J J A S O N D |

Brachysema latifolium
Broad-leaved Poison
S. | J F M A M J J A S O N D |

Daviesia colletioides

S. | J F M A M J J A S O N D |

Gompholobium confertum

S | J F M A M J J A S O N D |

Gastrolobium parviflorum
Box Poison
S. | J F M A M J J A S O N D |

Gastrolobium velutinum
Stirling Range Poison
S | J F M A M J J A S O N D |

Jacksonia elongata

S. | J F M A M J J A S O N D |

Bossiaea pressii

S. | J F M A M J J A S O N D |

Daviesia pachyphylla
Ouch Bush
S. | J F M A M J J A S O N D |

Bossiaea ornata
Broad- leaved Brown Pea
S. | J F M A M J J A S O N D |

FABACEAE - Pea family

Verticordia mitchelliana
Rapier Featherflower
S.W. ● J F M A M J J A S O N D

Verticordia grandiflora
J.S.W.Se. ● J F M A M J J A S O N D

Verticordia serrata
S. J F M A M J J A S O N D

Verticordia insignis
W.Se.S. J F M A M J J A S O N D

Veticordia longistylis
S. J F M A M J J A S O N D

Hypocalymma strictum
S.Se. J F M A M J J A S O N D

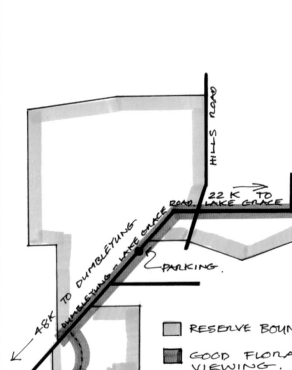
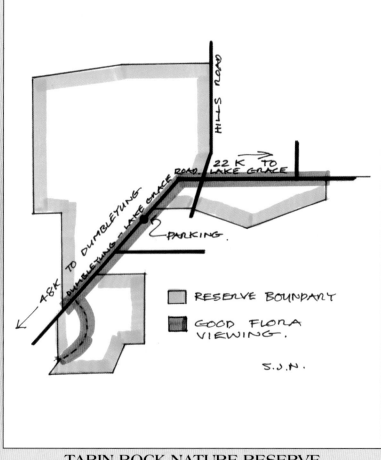

HILLS ROAD
22 K TO LAKE GRACE
ROAD. LAKE GRACE
48 K TO DUMBLEYUNG
DUMBLEYUNG - LAKE GRACE
PARKING.
RESERVE BOUNDARY
GOOD FLORA VIEWING.
S.J.N.

TARIN ROCK NATURE RESERVE

Siegfriedia darwinioides
S. J F M A M J J A S O N D

Drosera pallida
S.J.W. ● J F M A M J J A S O N D

Lechenaultia tubiflora.
Heath Lechenaultia
N.B.J.S. J F M A M J J A S O N D

Lechenaultia tubiflora yellow form.
Heath Lechenaultia
S. J F M A M J J A S O N D

Lechenaultia acutiloba
Wingless Lechenaultia
S. J F M A M J J A S O N D

Allocasuarina pinaster
Compass Bush
S. ● J F M A M J J A S O N D

Melaleuca fulgens subsp. *fulgens*
Scarlet Honeymyrtle
S.Se.W. ● | J F M A M J J A S O N D |

Melaleuca macronychia subsp. *macronychia*
S.Se. | J F M A M J J A S O N D |

Melaleuca spicigera
Se.S.W.J. | J F M A M J J A S O N D |

Melaleuca striata
S. | J F M A M J J A S O N D |

Melaleuca glaberrima
S.Se. | J F M A M J J A S O N D |

Melaleuca spathulata
J.S. ● | J F M A M J J A S O N D |

Melaleuca holosericea
Se.S.W.N. | J F M A M J J A S O N D |

Melaleuca suberosa
Cork Bark Honeymyrtle
S. ● | J F M A M J J A S O N D |

Melaleuca acerosa
N.B.W.S.J. | J F M A M J J A S O N D |

Melaleuca pungens
S. | J F M A M J J A S O N D |

Melaleuca sparsiflora
S. | J F M A M J J A S O N D |

Melaleuca diosmifolia
S. | J F M A M J J A S O N D |

Melaleuca bromelioides
S. | J F M A M J J A S O N D |

Melaleuca uncinata
Broom Bush
N.B.W.S.Se. ● | J F M A M J J A S O N D |

Melaleuca cuticularis
Saltwater Paperbark
B.S. ● | J F M A M J J A S O N D |

Melaleuca microphylla
B.S. ● | J F M A M J J A S O N D |

MELALEUCA

Banksia coccinea
Scarlet Banksia
S. JFMAMJJASOND

Banksia praemorsa
Cut leaf Banksia
S. JFMAMJJASOND ●

Banksia lemanniana
Lemann's Banksia
S JFMAMJJASOND

Banksia nutans subsp. *cernuella*
Nodding Banksia
S. JFMAMJJASOND

Banksia nutans subsp. *nutans*
S. JFMAMJJASOND ●

Banksia violacea
Violet Banksia
S.W. JFMAMJJASOND

Banksia speciosa
Showy Banksia
S. JFMAMJJASOND ●

Banksia laevigata subsp. *laevigata*
S JFMAMJJASOND

Banksia pilostylis
S. JFMAMJJASOND ●

BANKSIA

Banksia petiolaris
S.   J F M A M J J A S O N D

Banksia repens
Creeping Banksia
S.   J F M A M J J A S O N D   ●

Banksia gardneri subsp. *gardneri..*
S.   J F M A M J J A S O N D

Banksia baueri
Wooly Banksia
S.   J F M A M J J A S O N D   ●

Banksia occidentalis subsp. *occidentalis*
Red Swamp Banksia
S.J.   J F M A M J J A S O N D

Banksia pulchella
Teasel Banksia
S.   J F M A M J J A S O N D   ●

Banksia baxteri
Baxters Banksia
S   J F M A M J J A S O N D   ●

Banksia verticillata
Granite Banksia
S.   J F M A M J J A S O N D

Banksia seminuda
River Banksia
SK.J.   J F M A M J J A S O N D

BANKSIA

Hakea victoria
Royal Hakea
S. | J F M A M J J A S O N D

Calothamnus validus
Barrens Clawflower
S. | J F M A M J J A S O N D

Banksia oreophila
Western Mountain Banksia
S. | J F M A M J J A S O N D

Calothamnus pinifolius
Dense Clawflower
S. | J F M A M J J A S O N D

FITZGERALD RIVER NATIONAL PARK

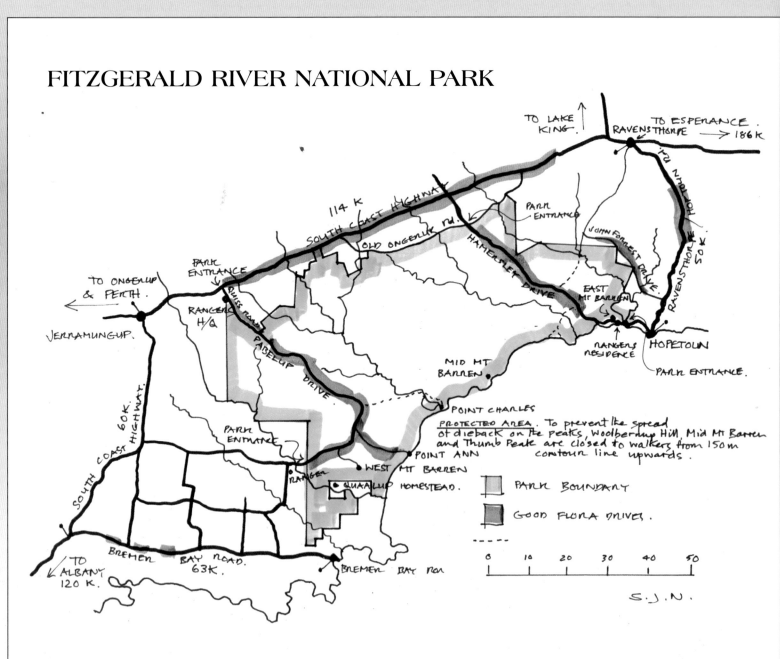

WESTERN AUSTRALIAN HERBARIUM

SCIENCE DIVISION

DEPARTMENT OF CONSERVATION AND LAND MANAGEMENT

Western Australia has an exceptionally rich array of biota, ecosystems and habitats. Over 12,500 kinds of vascular plant have been discovered in the state's 2.5 million square kilometres and more than 60% of the species are endemic to WA.

The south western part of the state is internationally recognised for its flowering plant diversity. This province alone is home to over 7,000 kinds of vascular plant.

The Western Australian Herbarium is an integral part of the Department of Conservation and Land Management. It is responsible for the description and documentation of Western Australia's botanical species diversity.

Biodiversity Collections

Our collections document the biodiversity of native and alien plants of Western Australia. 500,000 specimens include vascular plants, bryophytes, lichens, macro-fungi, marine and fresh water algae, as well as collections of micro-organisms of particular significance to conservation, agriculture and public health such as water moulds, Cyanobacteria, Actinobacteria.

Biosystematics

Biosystematics is the platform from which all biology advances. It provides the scientific basis for the understanding and ordering of taxonomic biodiversity by recognising species, classifying them into related groups and discerning their evolutionary history. The WA Herbarium includes a molecular genetics facility to extend its systematic research using the latest methods.

Current research elucidates the systematics of major plant groups, threatened species and those of economic importance as well as producing identification applications and manuals. The WA Herbarium also publishes the taxonomic journal *Nuytsia* and has

Banksia coccinea

produced a series of comprehensive regional floras. By establishing the names and relationships of plants and understanding their genetic systems and population dynamics we can:

- analyse evolutionary patterns and the delimitation and characterisation of species

- identify species at risk as well as targets for guided general and biodiscovery collecting

- highlight potential weed threats

Conservation Biology

Staff at the WA Herbarium survey, assess the conservation status and prepare area-based management plans for rare and threatened flora. This work is complemented by the *Threatened Flora Seed Centre* and integrated biological studies on critically endangered and other threatened plants.

To advise and assist in the conservation and recovery of these species we provide baseline data on:

- their ecology, life-history and seed biology

- the long term storage and collection of genetically representative seed

- genetic diversity patterns & mating systems

- re-introduction methodologies

FloraBase

Information on the Western Australian flora

What is it

FloraBase is a state-wide electronic flora information system integrating a range of authoritative biodiversity datasets into a readily accessible web site.

A FloraBase query can provide a list of plants growing in a particular bio-region, the most up-to-date name for a native plant or weed, or a short description with a range map and a representative image.

FloraBase draws on databases holding information on all 500,000 specimens in the state's herbarium and 17,000 names used now and in the past for WA plants. Together with 12,000 databased descriptions and computer-generated maps plus over 2,500 species images, it forms the most comprehensive online botanical resource in Australia.

Why is it important

Western Australia has a very diverse and unique flora. Almost one half of Australia's vascular flora grow in WA and nearly 80% are endemic to the state. Yet up to 20% are considered rare, threatened or have an uncertain conservation status.

With the growing community awareness of the needs for active conservation a reliable, up-to-date and readily accessible source of botanical information becomes essential.

All the resources and activities of the WA Herbarium are aimed at providing the latest scientific information with which to underpin the state's conservation effort.

Since 1998, FloraBase has become the primary method for the community to access this crucial botanical information.

Two examples of the FloraBase web site showing summary listing of plant names and further information available about each one, together with an example of a species page combining the available information for the rare Banksia brownii

Astroloma epacridis

New discoveries in the Western Australian flora are often made, necessitating an up-to-the-minute knowledge base for effective conservation and information dissemination on new or rare species

FloraBase
http://florabase.calm.wa.gov.au/

The user community

Since the launch of FloraBase onto the web in 1998, well over 2,000 regular users have registered for access at one of the content levels provided. The user profile developed during this time is a varied one.

While a proportion of users continues to be drawn from within the state conservation agencies where it has become an indispensable tool for anyone working with our native flora, as many users access it from around the globe.

FloraBase is used widely at all three levels of the education sector for school projects, local planting days or tertiary training and research.

Many users of FloraBase work or volunteer in various community conservation efforts or are members of community groups that coordinate revegetation and rehabilitation projects.

Scientists and botanical consultants around Australia and elsewhere in the world consult FloraBase for information that would otherwise take many hours to research and compile.

Sponsorship opportunities

FloraBase is modular in design and new content is constantly being added or prepared for release in forthcoming versions.

Projects developing modular content for FloraBase can be fast-tracked through appropriate sponsorship and include:

- more detailed descriptions of native species within particular taxonomic groups or regional ecosystems

- advanced information products for these target species groups such as interactive identification keys on CD, online as well as in print

- data capture for voucher specimens collected by community-based regional herbaria

- digitisation of existing regional flora management plans and interim recovery plans for publication on FloraBase

Professional scientists, community groups and a wide cross-section of the education sector access the online information resource FloraBase

- additional technical images illustrating for each species its appearance, life form and habitat, plus diagnostic features used in identification

- educational packages designed to help teachers and students achieve learning outcomes in the science, society and environment school curricula

Appropriate projects as suggested by potential sponsors will also be considered.

Information on the Western Australian flora

Contact details

Dr Neville Marchant
Western Australian Herbarium
Science Division
Department of Conservation
and Land Management

Locked Bag 104
Bentley Delivery Centre WA 6983

Phone: +61 8 9334 0500
Fax: +61 8 9334 0515
Email: nevillem@calm.wa.gov.au

Printed July 2001

REGIONAL HERBARIA NETWORK

What is it

The Regional Herbaria Network actively links regional community groups to the Western Australian Herbarium, creating nodes of local expertise collaboratively contributing to an expanding knowledge base on the WA flora.

The Regional Herbaria Network trains community groups to collect and document specimens of their local native and weedy plant species. Specimens collected are sent to Perth for identification and retention, while another remains in the local reference herbarium, cross-referenced by a unique number.

By compiling a local reference herbarium with specimens for which the currency of names is ensured, a regional group can create a local conservation resource.

The community can access further available information on local plant biodiversity through tools such as the FloraBase web site.

Why is it important

Local knowledge from regional herbarium groups allows us to gather comprehensive information about local floras.

Regional collectors know their local habitats and can access areas a visiting botanist may miss, providing year-round observations to document the life stages and ecological preferences of local species.

To date, some 14,000 well-documented specimens have been added jointly to the State Collection and the Regional Herbaria. Approximately 15 new species have been discovered and 676 conservation species vouchered, including 78 rare species.

The recent initiation of the Weed Information Network will support regional collectors making important contributions towards documenting the distribution of invasive species throughout the State.

Sponsorship opportunities

The Gordon Reid Foundation and the Natural Heritage Trust (NHT) initially funded the Regional Herbaria Network. Ongoing funding is required to implement or support the following range of initiatives:

- additional personnel to coordinate training and support for the 74 regional herbaria

- a customised series of regional training workshops focusing on identification and advanced collecting techniques

- additional infrastructural support for regional herbaria, including collecting and mounting materials and library resources

- web programming and content for FloraBase customised to provide improved support for each Regional Herbarium

- publication of user-friendly flora-style booklets for regional areas based on the work of regional herbaria

- an enhanced plant identification service for Regional Herbaria

The user community

The Regional Herbaria Network serves an estimated 700 regional volunteers and the broader community with 74 regional groups throughout the State.

Regional Herbaria are empowered to create and maintain a local reference collection providing access to up-to-date information about the region's flora.

Local government agencies, community groups and education facilities can all benefit from interacting with their local Regional Herbarium.

WEED INFORMATION NETWORK

What is it

The Weed Information Network (WIN) is a new Natural Heritage Trust supported initiative for the Western Australian Herbarium that over the next two years will develop a comprehensive weed-watch program and online information system for the State.

The Weed Information Network aims to become the primary authoritative information resource concerning the state's weed species. WIN is designed to integrate closely with the Regional Herbaria Network to focus local communities across the state on the problem of destructive invasive species.

Information products created as outcomes from the WIN project will be disseminated via the FloraBase web site.

Volunteers from local community conservation groups locate, document and collect voucher specimens of weeds for their regional herbarium

Why is it important

Weeds cause serious economic loss for agriculture and damage to the environment of natural areas. Good information is the first requirement for tackling the problems.

It is uncertain how many weed species occur in Western Australia because of poor information. Estimates range from 800 to over 1000 species. The WIN project is re-assessing the evidence.

Additionally, the actual identity of many weed species is uncertain due to a lack of expert identifications. WIN project botanists and cooperating international experts will review the identity of Western Australian weeds.

Specimens provide the physical evidence for the presence, distribution and identity of weeds. The WIN project will train and mobilise the many volunteers of the Regional Herbaria Network to collect good specimens with quality data from numerous locations across Western Australia.

The user community

A wide range of conservation practitioners in government and the community require authoritative information on the identification, propagation and control of invasive species.

Community groups forming the Regional Herbaria Network will be increasing their efforts in documenting local and regional weed species and will require the latest available information on weeds.

Sponsorship opportunities

The Weed Information Network (WIN) has recently obtained National Heritage Trust (NHT) funding establishing the basic frame-work. Supplementary projects may include:

- digital images of weed species for presentation in online and printed information products

- a customised series of regional training workshops focusing on weed identification and advanced collecting techniques

- an encouragement scheme to reward excellent community participation with practical equipment or botanical book gifts

Contact details

Dr Neville Marchant
Western Australian Herbarium
Science Division
Department of Conservation
and Land Management

Locked Bag 104
Bentley Delivery Centre WA 6983

Phone: +61 8 9334 0500
Fax: +61 8 9334 0515
Email: nevillem@calm.wa.gov.au

Printed July 2001

Mike   Euan   Janet,

Janet B.   Tim Bisseling —   Kiers

Poolbackal   Euan etal.

Stougard

Henridge   Broughh.   Kindorosa.
                    _____
Howiesa.   x   y.
_____

                              Halorrhagus

P53   Brachyremo.   (73)
 52   Isopoga
 5<   Unodn?

Biodiversity Information Systems

A range of authoritative biodiversity data is managed using sophisticated information technologies.

This allows us to communicate the results of our science to a wide range of users involved in conservation—ecologists, agriculturalists, educators, decision makers and members of the general community. Amongst these systems are the:

• Census of WA Plants – the authoritative listing of native and alien plant names

• specimen database – ca 500,000 plant specimen labels vouchering scientific knowledge

• plant descriptions – a short description for each vascular plant species in WA, complemented by comprehensive descriptions of all genera and families

• botanical library – the state's collection of botanical books, journals and archival material

• plant images – a collection of digital photographs based on vouchered specimens

• spatial data derived from specimen label details and other sources allowing us to analyse and visualise patterns in the distribution of the state's plants

• biological attributes of plants – a new information system concerned with issues such as salinity, disease and fire response.

FloraBase, a state-wide electronic flora, integrates all of these datasets into a single easy-to-use web site.

A FloraBase query can provide a list of plants growing in a particular bio-region, the most up-to-date name for a native plant or weed, or a short description with a range map and a representative image.

With the growing community awareness of the needs for active conservation, a reliable, up-to-date and readily accessible source of botanical information becomes essential.

Since 1998, FloraBase has become the primary method for the community to access this crucial botanical information.

Regional Herbaria Network

We support an extensive network of over 70 regional community groups to maintain local reference collections of duplicate specimens. The Regional Herbaria Network:

• trains volunteers to record detailed conservation data supported by voucher specimens

• contributes new documented specimens to the WA Herbarium collection and databases

• maintains the accuracy of identification and currency of naming for local reference collections by accessing FloraBase.

Pileanthus peduncularis

Weed Information Network

The Weed Information Network (WIN) is a new initiative backed by the Natural Heritage Trust that supports a comprehensive weed-watch program and provides an online weed information system for the state.

The Weed Information Network aims to become the primary authoritative information resource concerning the state's weed species. WIN is designed to integrate closely with the Regional Herbaria Network to focus local communities across the state on the problem of destructive invasive species.

Information products created as outcomes from the WIN project will be disseminated via the FloraBase web site.

Eremophila denticulata

Volunteer programs

The WA Herbarium has an exciting and popular program to mobilise the efforts of city-based and regional volunteers. Current activities by 60 metropolitan volunteers and ca 500 volunteers in regional areas of the state are:

- specimen processing

- image capture and storage

- plant identification

- provision of plant information to the tourism industry

- regional herbaria network coordination

- contributions to the electronic flora

- collection, identification and documentation of invasive species

- curation and identification in specialist plant groups

The WA Herbarium and associated regional herbaria together form a unique, dynamic, state-wide team which gathers, manages, researches and communicates information on the geography, systematics and biology of our unique and precious flora on behalf of all members of the Western Australian community.

In doing so it plays a vital role in a national and international network of herbaria and allied biodiversity conservation agencies.

Contact Details

Dr Neville Marchant
Western Australian Herbarium
Science Division
Department of Conservation
and Land Management

Locked Bag 104
Bentley Delivery Centre WA 6983
Australia

Phone: +61 8 9334 0500
Fax: +61 8 9334 0515
Email: nevillem@calm.wa.gov.au

Printed July 2001

uthern Mallee Shrublands and Heath

Pimelea physodes. red flowered form.
Qualup Bell.
S. ● J F M A M J J A S O N D

Lambertia inermis
Chittick
S. ● J F M A M J J A S O N D

Verticordia pityrhops
S. J F M A M J J A S O N D

Banksia caleyi
Caley's Banksia
S. J F M A M J J A S O N D

Acacia cedroides
Barrens Kindred Wattle
S. J F M A M J J A S O N D

Daviesia incrassata. subsp. *reversifolia*
S. J F M A M J J A S O N D

Daviesia striata
S. J F M A M J J A S O N D

View from East Mount Barren

Hibbertia mucronata
Prickly Hibbertia
S. J F M A M J J A S O N D

Darwinia vestita
Pom Pom Darwinea
S. J F M A M J J A S O N D

Melaleuca urceolaris
S. J F M A M J J A S O N D

Sphenotoma squarrosa
S. J F M A M J J A S O N D

Calothamnus macrocarpus
S. J F M A M J J A S O N D

Lasiopetalum bracteatum
Helena Velvet Bush
J.S. | J F M A M J J A S O N D

Lasiopetalum molle
Soft leaved Lasiopetalum
N.S. | J F M A M J J A S O N D

Chamelaucium magalopetalum
A Waxflower
S. | J F M A M J J A S O N D

Microcorys obovata
S.Se.W. | J F M A M J J A S O N D

Boronia crenulata
Aniseed Boronia
N.B.J.W.S.Se. | J F M A M J J A S O N D

Calytrix lechenaultii
S. | J F M A M J J A S O N D

Boronia crassifolia
S. | J F M A M J J A S O N D

Leptospermum sericeum
Silver Tea Tree.
S. | J F M A M J J A S O N D

Oligarrhena micrantha
S. | J F M A M J J A S O N D

Chamelaucium aorocladus
A Waxflower
S. | J F M A M J J A S O N D

Chamelaucium ciliatum
A Waxflower
S. | J F M A M J J A S O N D

Leptospermum spinescens
S. | J F M A M J J A S O N D

Ricinocarpos tuberculatus
Wedding Bush
S.Se. | J F M A M J J A S O N D

Synaphea acutiloba
Granite Synaphea
S.J.N.   J F M A M J J A S O N D

Stylidium schoenoides
Cow Kicks
J.S.   J F M A M J J A S O N D

Stylidium scandens
Climbing Triggerplant
J.S.   J F M A M J J A S O N D

Stylidium albomontis
S.   J F M A M J J A S O N D

Clematis microphylla
Old mans beard
S.Se.W.J.   J F M A M J J A S O N D

Stylidium rupestre
Rock Triggerplant
S.W.J.   J F M A M J J A S O N D

Stylidium breviscapum subsp. *erythrocalyx*
Boomerang Triggerplant
S.   J F M A M J J A S O N D

Stylidium pilosum
Silky Triggerplant
S.   J F M A M J J A S O N D

Conospermum leianthum
S.   J F M A M J J A S O N D

Lysinema conspicuum
S.   J F M A M J J A S O N D

Conospermum teretifolium
Spider Smokebush
S.   J F M A M J J A S O N D

Clematis pubescens
Old mans Beard
S.J.W.   J F M A M J J A S O N D

Anthoceris viscosa
Sticky tailflower
B.J.S.   J F M A M J J A S O N D

Conospermum caeruleum
S.   J F M A M J J A S O N D

Conospermum croniniae
Blue Smokebush
S.W.   J F M A M J J A S O N D

Conospermum bracteosum
S.   J F M A M J J A S O N D

Grevillea oligantha. robust form.

S. JFMAMJJASOND

Grevillea teretifolia. pink flowered form

N.W.Se.S. JFMAMJJASOND

Grevillea magnifica

S. JFMAMJJASOND

Grevillea decipiens

S.Se. JFMAMJJASOND

Conostylis bealiana
Yellow Trumpets
S. JFMAMJJASOND

Goodenia scapigera
White Goodenia ●
B.J.W.S. JFMAMJJASOND

Styphelia tenuiflora
Common Pinheath ●
N.B.J.S. JFMAMJJASOND

Scaevola striata

S. JFMAMJJASOND

Adenanthos sericeus subsp. sphalma

S. JFMAMJJASOND

Goodenia dyeri

S. JFMAMJJASOND

Eriostemon nodiflorus

S JFMAMJJASOND

Kunzea affinis

S.J. JFMAMJJASOND

Eremophila calorhabdos
Red Rod.
S. JFMAMJJASOND

Pityrodia exserta
Coastal Foxglove
S. JFMAMJJASOND

Astartea fascicularis

S. JFMAMJJASOND

Lachnostachys albicans

S.Se. JFMAMJJASOND

Flowers in full bloom, mid September in the Northern Kwongan sandplain, Kalbarri National Park.

Northern Mallee Shrubland and Heath

This zone, along with the related although separate Southern Mallee Shrubland and heath, is plant for plant the most biodiverse vegetation zone in the south west of Western Australia. This also makes it among the richest on the planet. The zone begins in the south either side of the northern prong of the Banksia and Eucalypt Woodland zone, and north of the wandoo woodland zone, running up to the top of the South West Botanical Province at Shark Bay, where it joins the Eremean Botanical Province. West is the Indian Ocean and east lies the Semi-arid Eucalypt Woodland zone.

Rainfall ranges from around 650mm at its southern limits, down to around 200mm at its northern limit, and inland rainfall drops to around 400mm at its central eastern edge reducing rapidly to the north.

The underlying geologies are interestingly and significantly varied and can often be observed as relief in the landscape. Here only the south eastern section of the zone is on the Yilgarn Shield where little relief in the landscape makes its edge less definable.

Biodiversity hot spots in the zone are underlain by particular geological formations including the lateralised sandstones, shales and siltstones of the Mt Leseuer area, where plant numbers and extremely localised endemism are among the highest and most spectacular in the State. These lateralised uplands are common throughout the zone and when observed in comparison to surrounding landscapes, plant numbers are noticeably greater. Other geological units also give rise to significant plant diversity, including the metamorphosed sandstone and gneisses of the Moora area, and the complex gneiss with granite intrusions and surrounding sandstones, siltstones and shales of the area to the north of Geraldton and around Kalbarri.

The duplex soils of the typical and extensive sandplains overlie foundations of marine and continental sandstones and siltstones that have seen less laterlisation than the uplands, although pockets do occur. The flora of these areas is also incredibly diverse, including many regional endemics.

The vegetation types making up the patterns in the zone are dominated by Kwongan heaths, also including massively diverse shrublands, with some woodlands mainly in valleys along the eastern edges.

The special significance is the range of considerably different Kwongan and scrubland types that occupy only slightly different soil types and topographical aspects.

The northern extremity of the zone is where a small number of very significant plants occur as the northernmost representatives of groups common in, and in many cases endemic to, the south west.

The array of plant diversity in almost all areas of this zone is at first subtle to the observer; when slight changes in soil types are noticed a corresponding abrupt and significant change in the component flora becomes apparent. When this phenomenon is added to the fact that enormous numbers of plant taxa occupy each soil type and that many different soil types exist, the tag "a coral reef out of water" seems a very appropriate description.

Banksia hookeriana
Hookers Banksia
N.Name J F M A M J J A S O N D ●

Banksia victoriae
Wooly Orange Banksia
N. J F M A M J J A S O N D ●

Banksia burdettii
Burdetts Banksia
N. J F M A M J J A S O N D ●

Banksia prionotes
Acorn Banksia
N.W.S. J F M A M J J A S O N D ●

Banksia leptophylla subsp. *leptophylla*
N. J F M A M J J A S O N D

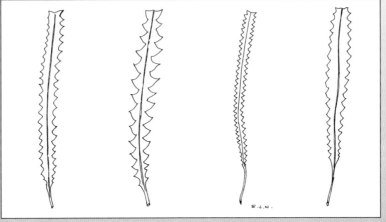
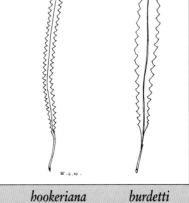

prionotes   victoriae   hookeriana   burdetti

Banksia micrantha
N. J F M A M J J A S O N D

Banksia scabrella
Burma Road Banksia
N. J F M A M J J A S O N D ●

Banksia incana
N.B. J F M A M J J A S O N D

Banksia grossa
N. J F M A M J J A S O N D

Banksia candolleana
Propeller Banksia
N. J F M A M J J A S O N D

Banksia chamaephyton
N. J F M A M J J A S O N D

Banksia sceptrum
Sceptre Banksia
N. J F M A M J J A S O N D ●

Banksia tricuspis
Lesueur Banksia
N. J F M A M J J A S O N D

Banksia lindleyana
Porcupine Banksia
N. J F M A M J J A S O N D

Banksia

Grevillea pinaster
N.W. JFMAMJJASOND

Grevillea dielsiana
Diels Grevillea
N. JFMAMJJASOND

Grevillea saccata
Pouched Grevillea
N. JFMAMJJASOND

Grevillea petrophiloides
Pink Pockers
N.W.Se. JFMAMJJASOND ●

Grevillea adpressa
N. JFMAMJJASOND

Grevillea amplexans
N. JFMAMJJASOND

Grevillea acrobotrya
N. JFMAMJJASOND

Grevillea uniformis
N. JFMAMJJASOND

Grevillea erynigoides
Curly Grevillea
N.W.Se. JFMAMJJASOND ●

Grevillea teretifolia. white form
N.W.Se.S. JFMAMJJASOND

Grevillea candelabroides
N. JFMAMJJASOND ●

Grevillea pilulifera
N. JFMAMJJASOND

Grevillea thysoides
N. JFMAMJJASOND

Grevillea dryandroides subsp. dryandroides
Phalanx Grevillea
N. JFMAMJJASOND

Grevillea synapheae subsp. pachyphylla
Catkin Grevillea
N. JFMAMJJASOND

Grevillea didymobotrya subsp.
didymobotrya
N. JFMAMJJASOND

GREVILLEA

Acacia stenoetera
N.

Acacia neurophylla subsp *erugata*
N.

Acacia restiacea
N.Se.

Acacia sphacelata subsp *sphacelata*
N.
Name

Acacia chrysocephai subsp. *obovata*
N.

Acacia blakely
N.

Acacia latipes
N.

Acacia coolgardiensis subsp. *coolgardiensis*
N

Monotaxis grandiflora
Diamond of the desert
N.M.

Jacksonia nutans
N.

Glischrocaryon flavescens
N.

Dryandra carlinoides
Pink Dryandra
N.

Verticordia monadelpha white form
Woolly Featherflower
N

Thryptomene baekeacea
N.

Keraudrenia integrifolia
N.W.S.Se.

Thryptomene hyporhytis
N.

Conospermum nervosum
N. [JFMAMJJASOND]

Conospermum acerosum
Needle-leaved Smokebush
N.B. [JFMAMJJASOND]

Conospermum incurvum
Plume Smokebush
N. [JFMAMJJASOND]

Conospermum crassinervium
Summer Smokebush
N. [JFMAMJJASOND]

Eriostemon spicatus
Pepper and Salt
N.B.J.Se. [JFMAMJJASOND]

Scholtzia uberiflora
N. [JFMAMJJASOND]

Synaphea polymorpha
N.J.S. [JFMAMJJASOND]

Lambertia multiflora red form
Many-flowered Honeysuckle
N.B.J. [JFMAMJJASOND]

Cyaniculata deformis
Blue fairy Orchid
N.B.J.W.S.Se. [JFMAMJJASOND]

Thelymitra antennifera
Vanilla Orchid
N.B.J.W.S.Se.M. [JFMAMJJASOND]

Caladenia flava Subsp. *flava*
Cowslip Orchid
N.B.J.W.S.Se. [JFMAMJJASOND]

Diplopeltis huegeli var *lehmanii*
Pepper Flower
N [JFMAMJJASOND]

Eriochilus dilatatus subsp. *undulatus*
Crinkle Leaf Bunny Orchid
N.W.S.Se. [JFMAMJJASOND]

Prasophyllum plumaeforme
Dainty Leek Orchid
N.B.J.S. [JFMAMJJASOND]

Stylidium crossocephalum
Posy Triggerplant
N.B. [JFMAMJJASOND]

Stylidium elongatum
Tall Triggerplant
N. [JFMAMJJASOND]

87

Hakea costata
Ribbed Hakea
N.B. `J F M A M J J A S O N D` ●

Hakea prostrata (inland plants erect)
Harsh hakea
N.Se.W.S. `J F M A M J J A S O N D` ●

Hakea conchifolia
Shell-leaved Hakea
N. `J F M A M J J A S O N D` ●

Hakea gilbertii

N.W. `J F M A M J J A S O N D`

Hakea lissocarpha
Honey Bush
B.J.W.Se.S. `J F M A M J J A S O N D`

Calothamnus quadrifidus
One sided Bottlebrush
N.W.J.B.S.Se. `J F M A M J J A S O N D`

Calothamnus torulosus

N. `J F M A M J J A S O N D`

Calothamnus blepharospermus

N. `J F M A M J J A S O N D`

Calothamnus oldfeldii

N.Se.M. `J F M A M J J A S O N D`

Eucalyptus macrocarpa
Mottlecah
N.W. J F M A M J J A S O N D

Eucalyptus macrocarpa -x *pyriformis* Hybrid
N. J F M A M J J A S O N D

Eucalyptus pyriformis red form
Dowerin Rose
N.W. J F M A M J J A S O N D

Eucalyptus rhodantha
W. J F M A M J J A S O N D

Eucalyptus erythrocorys
Illyarrie
N. J F M A M J J A S O N D

Eucalyptus camaldulensis
Red River Gum
N.Se.M. J F M A M J J A S O N D

Eucalyptus oldfiedii
N.Se.M. J F M A M J J A S O N D

Eucalyptus todtiana
Coastal Blackbutt
B.N. J F M A M J J A S O N D

Eucalyptus pendens
Badgingarra Mallee
N. J F M A M J J A S O N D

Eucalyptus

Astroloma xerophyllum
N.B.W. J F M A M J J A S O N D

Astroloma serratifolium
Kondrung
N. J F M A M J J A S O N D

Astroloma microdonta
Sandplain Cranberry
N. J F M A M J J A S O N D

Astroloma prostratum
N. J F M A M J J A S O N D

Conostylis canteriata
N. J F M A M J J A S O N D

Conostylis robusta
Golden Conostylis
N.B.J.W. J F M A M J J A S O N D

Conostephium pendulum
Pearl Flower
B.N.J. J F M A M J J A S O N D

Astroloma stomarrhena
Red Swamp Cranberry
N. J F M A M J J A S O N D

Boronia cymosa
Granite Boronia
N.B.J. J F M A M J J A S O N D

Scholtzia uberiflora
N. J F M A M J J A S O N D

Calectasia grandiflora
Blue Tinsel Lily
N.B.W.Se.S. J F M A M J J A S O N D

Eremaea violacea
Violet Eremaea
N. J F M A M J J A S O N D

Calytrix flavescens
Summer Starflower
N.B.W J F M A M J J A S O N D

Calytrix brevifolia
N. J F M A M J J A S O N D

Hybanthus calycinus
Wild Violet
N.B. J F M A M J J A S O N D

Hypocalymma xanthopetulum
N.B. J F M A M J J A S O N D

Anigozanthos pulcherrimus
Yellow Kangaroo Paw
N.　　　J F M A M J J A S O N D

Anigozanthus humilis subsp. *humilis*
Common Catspaw
N.W.S.　　　J F M A M J J A S O N D

Macropida fuliginosa
Black Kangaroo Paw
N.　　　J F M A M J J A S O N D

Diplolaena grandiflora
Tamala Rose
N.　　　J F M A M J J A S O N D

Cyanostegia corifolia

N.W.　　　J F M A M J J A S O N D

Darwinia speciosa

N.　　　J F M A M J J A S O N D

Darwinia virescens
Murchison Darwinea
N.　　　J F M A M J J A S O N D

Diplolaena ferruginea

N.　　　J F M A M J J A S O N D

Dryandra speciosa
Shaggy Dryandra
N.Se.　　　J F M A M J J A S O N D

Dryandra ashbyi

N.　　　J F M A M J J A S O N D

Anthocerias littorea
Yellow Tailflower
N.B.J.S.　　　J F M A M J J A S O N D

Petrophile brevifolia

N.　　　J F M A M J J A S O N D

Petrophile linearis
Pixie Mops
N.B.J.　　　J F M A M J J A S O N D

Petrophile inconspicua

N.　　　J F M A M J J A S O N D

Petrophile chrysantha

N.　　　J F M A M J J A S O N D

Corynanthera flava

N.　　　J F M A M J J A S O N D

Chamelaucium uncinatum
Geraldton Wax
N.B. J F M A M J J A S O N D

Pityrodia bartlingii
Wooly Dragon
N.B.W.S.Se. J F M A M J J A S O N D

Physopsis lachnostachya
N. J F M A M J J A S O N D

Scaevola phlebopetala
Velvet Fanflower
N.B.J.W.S. J F M A M J J A S O N D

Johnsonia pubescens
Pipe Lily
N.B. J F M A M J J A S O N D

Hibbertia pungens
N. J F M A M J J A S O N D

Beaufortia elegans
N. J F M A M J J A S O N D

Beaufortia aestiva
N. J F M A M J J A S O N D

Lasiopetalum drummondii
N. J F M A M J J A S O N D

Jacksonia rigida
N. J F M A M J J A S O N D

Guichenotia ledifolia
N.B.S. J F M A M J J A S O N D

Dampiera lindleyi
N. J F M A M J J A S O N D

Conospermum boreale subsp. *boreale*
N. J F M A M J J A S O N D

Comesperma scoparium
Broom Milkwort
N.W.Se. J F M A M J J A S O N D

Tersonia cyathiflora
Button Creeper
N.B. J F M A M J J A S O N D

Ecdeiocolea monostachya
N.W.Se. J F M A M J J A S O N D

Grevillea eriostachya
Flame Grevillea
N.B.W.Se. ● J F M A M J J A S O N D

Grevillea polybotrya
N.Se. ● J F M A M J J A S O N D

Grevillea annulifera
Prickly Plume Grevillea
N. ● J F M A M J J A S O N D

Grevillea leucopteris
Old Socks
N. ● J F M A M J J A S O N D

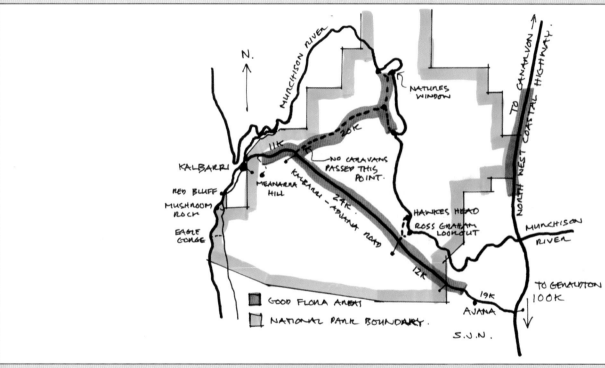

KALBARRI NATIONAL PARK

Geleznowia verrucosa

N.Se. J F M A M J J A S O N D

Pentaptilon careyi

N. ● J F M A M J J A S O N D

Pityrodia oldfieldi
Oldfields Foxglove
N.B.J.W.Se. ● J F M A M J J A S O N D

Hibbertia glomerosa
Guinea-flower
N.W.Se. ● J F M A M J J A S O N D

93

Verticordia etheliana

N.   J F M A M J J A S O N D

Verticordia grandis
Scarlet Featherflower

N.   J F M A M J J A S O N D ●

Verticordia muelleriana

N.W.   J F M A M J J A S O N D

Verticordia nobilis

N.B.   J F M A M J J A S O N D ●

Verticordia oculata

N.   J F M A M J J A S O N D

Verticordia chrysostachya

N.   J F M A M J J A S O N D

Verticordia oculata

N.W.Se.   J F M A M J J A S O N D

Verticordia monadelpha
KALBARRI NATIONAL PARK

Verticordia dichroma var *dichroma*

N.   J F M A M J J A S O N D

Verticordia eriocephala
Common Couliflower
N.W.Se.   J F M A M J J A S O N D ●

Verticordia chrysostachya subsp. *palida*

N.   J F M A M J J A S O N D

Verticordia cooloomia

N   J F M A M J J A S O N D

Verticordia muelleriana X *chrysotachya palida* hybrid

N   J F M A M J J A S O N D

VERTICORDIAS · FEATHERFLOWERS

Sphaerolobium pulchellum

N. J F M A M J J A S O N D

Gastrolobium bidens
Hill River Poison ●

N. J F M A M J J A S O N D

Gastrolobium laytoni
Breelya

N J F M A M J J A S O N D

Gastrolobium phyllosum ●

N. J F M A M J J A S O N D

Daviesa podaphylla

N. J F M A M J J A S O N D

Daviesia triflora

N J F M A M J J A S O N D

Daviesia nudiflora

N. J F M A M J J A S O N D

Daviesia divaricata
Marno

N. J F M A M J J A S O N D

Gastrolobium bennettsianum

N. J F M A M J J A S O N D

Nemcia capitata
Bacon & Eggs ●

N. J F M A M J J A S O N D

Mirbelia spinosum

N. J F M A M J J A S O N D

Mirbelia depressa

N. J F M A M J J A S O N D

Euchilopsis linearis
Swamp Pea
Name J F M A M J J A S O N D

Gompholobium knightianum
Handsome Wedge Pea

N. J F M A M J J A S O N D

Isotropis cuneifolia
Common lamb Poison

N. J F M A M J J A S O N D

Jacksonia floribunda ●
Holly Pea

N. J F M A M J J A S O N D

FABACEAE - Pea family

Daviesia epiphylla
Staghorn Bush
N. | J | F | M | A | M | J | J | A | S | O | N | D |

Kennedia eximia
N. | J | F | M | A | M | J | J | A | S | O | N | D |

Gastrolobium polystachyum
Horned Poison
N. | J | F | M | A | M | J | J | A | S | O | N | D |

Jacksonia ulicina
N. | J | F | M | A | M | J | J | A | S | O | N | D |

Jacksonia velutina
N. | J | F | M | A | M | J | J | A | S | O | N | D |

Mirbelia spinosa
N.B.W.Se.S. | J | F | M | A | M | J | J | A | S | O | N | D |

Sphaerolobium macranthum
Globe Pea
N.B. | J | F | M | A | M | J | J | A | S | O | N | D |

Petrophile plumosa
N. | J | F | M | A | M | J | J | A | S | O | N | D |

Melaleuca holosericea
N.W.Se.S. | J | F | M | A | M | J | J | A | S | O | N | D |

Hemigenia macrantha
N. | J | F | M | A | M | J | J | A | S | O | N | D |

Lechenaultia hirsuta
Hairy Lechenaultia
N. | J | F | M | A | M | J | J | A | S | O | N | D |

Lechenaultia linarioides
Yellow Lechenaultia
N.B. | J | F | M | A | M | J | J | A | S | O | N | D |

Lechenaultia macrantha
Wreath Lechenaultia
N. | J | F | M | A | M | J | J | A | S | O | N | D |

Pileanthus filifolius
Summer Coppercups
N.Se. | J | F | M | A | M | J | J | A | S | O | N | D |

Pileanthus peduncularis
Coppercups
N.Se. | J | F | M | A | M | J | J | A | S | O | N | D |

Hemiandra linearis
Speckled Snakebush
N.W. | J | F | M | A | M | J | J | A | S | O | N | D |

Open Mallee woodland due east of Hyden.

Semi-arid Eucalypt Woodland

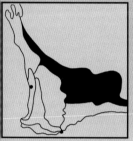

The semi arid woodland zone is one of the great contrasts as well has having some similarities to the wandoo woodland zone to the west.

Separating the Wandoo and the Semi arid Eucalypt Woodland zones is important as there are obvious differences between the two. The principal difference is in composition. The Wandoo shares its zone with only a few other woodland trees, while the semi arid woodland zone contains many different eucalyptus species occurring as a matrix of woodlands interspersed with other vegetation types.

The northern and eastern boundaries of this zone run into a botanical area known as the South West Interzone, where a remarkable feature is eucalypt woodlands dominating the landscape in a rainfall of less than 200 mm per year. The geology of this zone is wholly underlain by the massive granite Yilgarn Shield, with common greenstone intrusions through the granite, featuring rich mineralisation with significant gold and nickel resources.

Soil types derived from these geological units are complex and usually rich in nutrients compared to Kwongan soils. The other significant landscape feature is the expansive broad valleys, often occupied by naturally occurring salt lake systems. These systems, visible on maps as chains of lakes, are the remains of great river systems that flowed north west across Gondwana, off what is now Antarctica when it was joined to Australia, entering the Indian Ocean some 50 or so kilometres west of the present shoreline.Some also drained south east into the Nullarbor Basin.

These systems assisted the building of deeper loamy soils in the valleys on which the woodlands now occur, or perhaps more significantly once occurred, for these are also the soils selected for broad scale clearing for the development of the so called "wheatbelt". The majority has been cleared within the last century, leaving a degradation legacy that will take enormous commitment, knowledge and resources to restore.

These woodlands are extremely rich in type as well as component

diversity, offering some interesting features of just how complex the patterns can be.

One fine example demonstrates the concept that the boundary between the Wandoo and this zone is fuzzy and not as obvious as others. Wandoo (*Eucalyptus wandoo*) and its very similar species, inland wandoo, (*Eucalyptus capilloa*). While these woodland trees remain outwardly similar in appearance, the ecological place and the sites they occupy are quite different. Inland Wandoo occurs in a wide and very scattered distribution in the semi arid woodland zone mainly on decomposing breakaway systems, so it doesnt occur like its western relation as a widespread and common type.

This remarkable matrix of woodland occurs in most cases, exclusively due to subtleties in soil type, creating some of the most spectacular mosaics of woodlands and other systems, including:

- Salmon gum *E salmonophloia)* woodlands
- The gimlets, often as abrupt pure woodlands, (*E salubris, E campaspe, E glauca, E ravida and E diptera,)*
- Blackbutt woodlands of *E dundasii, E corrugata* and *E lesoueffi,* coral gum *(E torquata)* and several unnamed species,
- Redwood woodlands, *E transcontinentalis,* and several related and unnamed species,
- Morrel woodlands (*E longicornis* and *E melanoxylon),*
- Salt gum woodlands (*E salicola)* around salt lakes,
- Granite outcrop communities,
- Extensive Kwongan communities,
- Salt lake margin saltbush heaths,
- Melalueca thickets on water gaining sites, and
- Grassland and herbfield understoreys.

Visually striking features of this zone are the dramatic "red- or orange-ness" of some of the soils, and the clear differences in the colours of the trees in each woodland type. Perhaps the most remarkable aspect about the importance of this zone is that new species of woodland trees, along with other plants, are still being discovered.

Banksia ashbyi
Ashby's Banksia
N.Se. ● JFMAMJJASOND

Banksia benthamiana
Bentham's Banksia
Se. JFMAMJJASOND

Banksia audax
Se. JFMAMJJASOND

Banksia laevigata subsp. *fuscolutea*
Tennis Ball Banksia
Se. JFMAMJJASOND

Amyema gibberula
Mistletoe
W.Se.Name ● JFMAMJJASOND

Petrophile trifida
Se. JFMAMJJASOND

Petrophile incurvata
W.Se. JFMAMJJASOND

Petrophile ericifolia
W.S.Se. ● JFMAMJJASOND

Petrophile shuttleworthiana
N.Se. JFMAMJJASOND

Isopogon scabriusculus
W.Se. JFMAMJJASOND

Dryandra vestita
Summer Dryandra
Se. JFMAMJJASOND

Melaleuca ciliosa
Se. JFMAMJJASOND

Melaleuca fulgens
Scarlet Honeymyrtle ●
S.Se.W. JFMAMJJASOND

Conospermum brownii
Blue-eyed Smokebush ●
N.W.Se. JFMAMJJASOND

Conospermum croniniae
Se. ● JFMAMJJASOND

Dampiera wellsiana
Wells Dampiera
Se JFMAMJJASOND

Hakea multilineata
Grass-leaved Hakea
W.Se. J F M A M J J A S O N D ●

Hakea francisiana
Emu Tree
N.Se. J F M A M J J A S O N D ●

Hakea baxteri
Fan Hakea
Se J F M A M J J A S O N D ●

Hakea erecta
N.Se.M. J F M A M J J A S O N D

Hakea roei
Se. J F M A M J J A S O N D ●

Hakea platysperma
Cricket Ball Hakea
N.Se.W. J F M A M J J A S O N D ●

Hakea scoparia
W.Se.N. J F M A M J J A S O N D

Hakea bucculenta
Red Pokers
N.Se. J F M A M J J A S O N D ●

Hakea preissii
Needle Tree
N.Se.S. J F M A M J J A S O N D ●

HAKEA

Melaleuca elliptica
Granite Honeymyrtle
S.Se. J F M A M J J A S O N D

Hakea verrucosa
Wheel Hakea
Se. J F M A M J J A S O N D

Melaleuca cardiophylla subsp. *longistaminea*
Tangling Melaleuca
W.N.Se.S. J F M A M J J A S O N D

Melaleuca conothamnoides

N.W.Se J F M A M J J A S O N D

Melaleuca cordata

W.Se. J F M A M J J A S O N D

Melaleuca filifolia
Wiry Honeymyrtle
Se. J F M A M J J A S O N D

Actinostrobus arenarius
Sansplain Cypress
N.W.Se. J F M A M J J A S O N D

Allocasuarina corniculata

Se. J F M A M J J A S O N D

Allocasuarina tortiramula
Twisted Sheoak
Se. J F M A M J J A S O N D

Gyrostemon racemiger

N.B.S.W.Se. J F M A M J J A S O N D

Santalum murryanum
Bitter Quandong
S.W.Se.N. J F M A M J J A S O N D

Veticordia picta
Painted Featherflower
N.B.W.Se. J F M A M J J A S O N D

Verticordia roei
Roe's Featherflower
S.Se.W. J F M A M J J A S O N D

Verticordia pennigera

Se J F M A M J J A S O N D

Dampiera spicigera
Spiked Dampiera
N.Se. J F M A M J J A S O N D

Halgania andromedifolia

S.Se.W. J F M A M J J A S O N D

Astroloma serratifolium var. *horridulum*
Kondrung
N.Se. ● | J F M A M J J A S O N D |

Brachyloma concolor
Se. | J F M A M J J A S O N D |

Drummondita hassellii var. *hassellii*
Se. | J F M A M J J A S O N D |

Trymalium myrtillus subsp. *pungens*
Se. | J F M A M J J A S O N D |

Acacia chrysocephala
Se. | J F M A M J J A S O N D |

Acacia coolgardiensis subsp. *coolgardiensis*
Spinifex Wattle
Se. | J F M A M J J A S O N D |

Acacia inaequiloba
Se. | J F M A M J J A S O N D |

Acacia acuaria
Se. | J F M A M J J A S O N D |

Acacia verricula
Se. | J F M A M J J A S O N D |

Acacia morraii subsp *recurvistiula*
Se.
Name | J F M A M J J A S O N D |

Acacia mimica var. *mimica*
Se. | J F M A M J J A S O N D |

Acacia rosseii
Se. | J F M A M J J A S O N D |

Acacia acuminata
Jam ●
Se.W.N.S.M. | J F M A M J J A S O N D |

Acacia longiphyllodinea
Long-leaved Wattle ●
Se | J F M A M J J A S O N D |

Acacia barbinervis
Se | J F M A M J J A S O N D |

Acacia sp.
N.W.Se. ● | J F M A M J J A S O N D |

Semi-arid Eucalypt Woodland

Triodia danthonides

Se.

Rulingia densifolia

N.Se.

Persoonia micrcarpa

J.W.S.Se.

Pityrodia teckiana

Se.

Caladenia saccharata
Sugar Orchid
W.S.Se.

Caledenia longicauda

W.Se.

Persoonia rufiflora

N.B.S.Se.

Anthocercis littorea
Yellow Tailflower
N.B.S.Se.

Grevillea pauciflora subsp.*psilophylla*

Se.

Grevillea huegeli

N.B.J.S.Se.

Gompholobium scabrum

N.B.J.W.S.Se.

Disphyma crassifolium
Round leaved Pigface
Se.S.W.N.M.

Calytrix angulata
Yellow Starflower
B.J.W.Se.

Jacksonia acicularus

Se.

Guichenotia micrantha
Large-flowered Guichenotia
N.B.S.W.Se.

Boronia ternata

Se.

102

Eucalyptus grossa
Coarse-leaved Mallee
Se. | J F M A M J J A S O N D |

Eucalyptus torquata
Coral Gum
Se. | J F M A M J J A S O N D |

Eucalyptus stowardii
Fluted horn Mallee
Se. | J F M A M J J A S O N D | ●

Eucalyptus eremophila subsp. *eremophila*
Sand Mallee
Se. | J F M A M J J A S O N D | ●

Eucalyptus woodwardi
Se. | J F M A M J J A S O N D |

Eucalyptus stoatei
Scarlet Pear Gum.
Se. | J F M A M J J A S O N D |

Eucalyptus incrassata
Ridge-fruited Mallee
Se.S. | J F M A M J J A S O N D | ●

Eucalyptus loxophleba subsp. *loxophleba*
York gum
Se.W.B.N. | J F M A M J J A S O N D | ●

Eucalyptus salubris var. *salubris*
Gimlet
Se. | J F M A M J J A S O N D | ●

Eucalyptus salmonophloia
Salmon gum
W.Se. | J F M A M J J A S O N D | ●

Eucalyptus melanoxylon
Black morrel
Se. | J F M A M J J A S O N D |

EUCALYPTUS

Balaustion microphyllum
Bush Pomegranate
Se. J F M A M J J A S O N D

Alyogyne hakeifolia
Red centred Hibiscus
N.B.S.Se. J F M A M J J A S O N D

Grevillea shuttleworthiana subsp. *obvata* ●
Se. J F M A M J J A S O N D

Scaevola thesioides
Se J F M A M J J A S O N D

Waitzia acuminata
Orange Immortelle ●
Se.M. J F M A M J J A S O N D

Dioscorea hastifolia
Warrine ●
N.B.J.Se. J F M A M J J A S O N D

Leptosema daviesiodes
Upside down Pea
N.Se.M. J F M A M J J A S O N D

Pityrodia terminalis
Native Foxglove
N.B.Se.M. J F M A M J J A S O N D

Anthotium rubriflorum
Red Anthotium ●
S.Se. J F M A M J J A S O N D

Dioscorea hastifolia
J F M A M J J A O N D

Chamaexeros fimbriata ●
Se. J F M A M J J A S O N D

Prostanthera serpyllifolia
Small leaf Mintbush
J.S.Se. J F M A M J J A S O N D

Halosarcia doleiformis
Samphire ●
Se.W.S.N.M. J F M A M J J A S O N D

Gnephosis tenuissima
Se.W.N. J F M A M J J A S O N D

Darwinea purpurea
Rose Darwinea
Se J F M A M J J A S O N D

Phebalium ambigua
Se.M. J F M A M J J A S O N D

Grevillea armigera
Prickly Toothbrush
Se. J F M A M J J A S O N D

Grevillea granulosa
Se.N.M. J F M A M J J A S O N D

Grevillea uncinulata
Hook-leaved Grevillea
N.Se. J F M A M J J A S O N D

Grevillea asparagoides
Se. J F M A M J J A S O N D

Grevillea paradoxa
Bottlebrush Grevillea
N.Se.M. J F M A M J J A S O N D

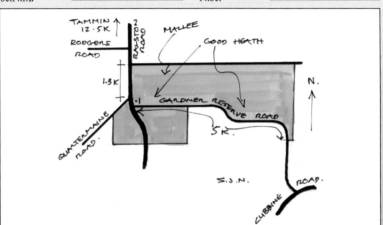

CHARLES GARDNER NATURE RESERVE

Grevillea wittweri
Se J F M A M J J A S O N D

Grevillea tetrapleura

Se J F M A M J J A S O N D

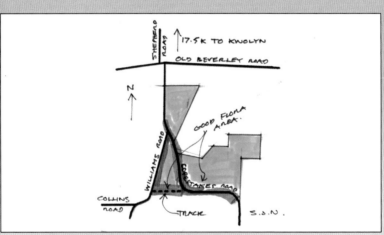

BOOLANOOLING NATURE RESERVE

Grevillea levis
Se.M. J F M A M J J A S O N D

Grevillea commutata
Se. J F M A M J J A S O N D

Grevillea apiciloba subsp. *digitata*
Se. J F M A M J J A S O N D

Grevillea hookeriana simple leaf form
Black Toothbrushes
S.W.N.Se. J F M A M J J A S O N D

Grevillea cagiana robust form
Se. J F M A M J J A S O N D

GREVILLEA

Mullamullas, Everlastings and the weed Ruby Dock in full bloom north of Mount Magnet.

The Mulga

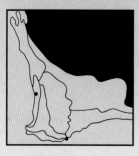

The mulga zone lies outside the South West Botanical Province, although included in this guide because of its proximity to the south west and its comprehensively different set of vegetation systems.

The dominant plant of this zone, the mulga, is an acacia or wattle (*Acacia aneura*), occurring as several differing forms across arid Australia.

The edge of the mulga zone marks the change from eucalypt-dominated botanical systems of the south west to that dominated by acacias, with fields of grasses, annual plants and scattered shrubs as wide uncluttered understoreys. However, as with the south western systems, mosaic of vegetation types is evident.

The mulga plant typically grows along watercourses down which heavy rains drain a little less rapidly than the surrounding landscape, away into larger rivers, claypans and salt lakes. The climate of this zone is typically arid with rainfall less than 200 mm per year, occurring spasmodically during any of the seasons, rainfall being the dominant driving force for the ebb and flow of the arid zone vegetation systems. Winters are frequently dry and cold with night temperatures occasionally dropping below zero.

The makeup of the landscape across the zone is one of broad flat plains, sandy mini deserts, rough rocky rises, breakaways, granite outcrops, watercourses as creeks and rivers, salt lakes and clay pans.

This zone is where the popular 'fields of colour' wildflower experiences are celebrated when sufficient rains create the opportunities for carpets of everlastings to appear. Everlastings are mostly members of the daisy family Asteracae, occurring as several genera.

This landscape supports a great diversity of plants mostly at their best after rain. Great diversities of emu bushes (*Eremophila sp*) and mulla mulla (*Ptilotus sp*) occur, often alongside the famous Desert Pea (*Swainsona formosa*) with its glossy crimson flowers and black or maroon centre boss.

Spinifex grasslands of *Triodia* create one of the great Australian scenes across wide sandy plains, mostly with the landscape to themselves.

Creek and river systems support a line of woodland fringes where red river gums (*Eucalyptus camaldulensis*) form the dominant vegetation type.

This zone provides a significant contextual factor in outlining the way in which the natural systems of Australia, in this case the south west of Western Australia, have evolved quite separately given the climatic sequence with which they have been provided. A noticeable legacy of this is the readily observable fact that many of the plants of the mulga have few relations in the South West Botanical Province.

Goodenia berardiana
N.S.Se.M. JFMAMJJASOND

Olearia muelleri
Goldfields Daisy
N.S.Se.M. JFMAMJJASOND

Senecio magnificus
Tall Yellow Top
Se.M. JFMAMJJASOND

Rhodanthe chlorocephala subsp *.rosea*
N.B.W.Se.M. JFMAMJJASOND

Brunonia australis
Native Cornflower
N.Se.M. JFMAMJJASOND

Cephalipterum drummondii
Pom Pom Head
N.M.Se. JFMAMJJASOND

Podolepis canescens
Bright Podolepis
N.B.W.J.Se.S.M. JFMAMJJASOND

Podolepis gardneri
N.M.Se. JFMAMJJASOND

Rhodanthe chlorocephala subsp. *splendida*
N.M.Se. JFMAMJJASOND

Podotheca gnaphalioides
Golden Longheads
N.B.J.W.Se.M. JFMAMJJASOND

Brachyscome iberidifolia
Native Daisy
N.B.J.S.Se.M. JFMAMJJASOND

Shoenia cassiniana
N.Se.M. JFMAMJJASOND

Waitzia suaveolens
Fragrant Waitzia
N.B.W.Se.M. JFMAMJJASOND

Lawrencella davenportii
Sticky Everlasting
N.S.W.Se.M. JFMAMJJASOND

Rhodanthe chlorocephala subsp. *splendida*
Splendid Everlastings
N.M.Se. JFMAMJJASOND

Leptosema sp.
N.M. JFMAMJJASOND

Eremophila foliosissima
M.

Eremophila glabra
Tar Bush
N.B.Se.M.

Eremophila oldfieldii
Pixie Bush
M.

Eremophila latrobei
Warty Fuchsia Bush
M.

Eremophila forrestii
Wilcox Bush
M.

Eremophila fraseri
Turpentine Bush
M.

Ptilotus exaltatus
Tall Mulla Mulla
M.

Ptilotus chamaecladus
M.

Ptilotus macrocephalus
Featherheads
M.

Ptilotus obovatus
Cotton Bush
M.

Pityrodia atriplicina
N.M.

Newcastelia chrysophylla
Golden Lambstail
Se.M.

Swainsona colutoides
Bladder Vetch
M.

Swainsonia sp.
M.

Eriostemon brucei
Noolburra
M.

Pittosporum phylliraeoides
M.Se.

Maireana carnosa
Cottony Bluebush
N.Se.M.   J F M A M J J A S O N D

Senna glutinosa
Sticky Caesia
N.Se.M.   J F M A M J J A S O N D

Maireana georgei
Golden Bluebush
Se.M.   J F M A M J J A S O N D

Dodonea stenozyga
S.Se.M.   J F M A M J J A S O N D

Callitris glaucophylla
White Cypress Pine
Se. M.   J F M A M J J A S O N D

Alyxia buxifolia
Se.M.   J F M A M J J A S O N D

Solanum lasiophyllum
Flannel Bush
M.   J F M A M J J A S O N D

Swainsona formosa
N.Se.M.   J F M A M J J A S O N D

Indigofera georgei
Bovine Indigo
M.   J F M A M J J A S O N D

Phebalium filifolium
Slender Phebalium
Se.M.   J F M A M J J A S O N D

Darwinia masonii
Nigham Bell
M.   J F M A M J J A S O N D

Velleia rosea
Pink Velleia
N.Se.M.   J F M A M J J A S O N D

Hakea suberea
Cork Tree
M.   J F M A M J J A S O N D

Santalum spicatum
Sandlewood
M.Se.   J F M A M J J A S O N D

Acacia quadrimarginea
Granite Wattle
M.   J F M A M J J A S O N D

Grevillea acacioides
M.   J F M A M J J A S O N D

ABORIGINAL NAMES

Due to the fact that there was no written language for the various Aboriginal tribal groups throughout Australia, it was difficult for the early settlers to establish precise vernacular names for the various trees and plants they encountered. This problem was compounded by the fact that there are several tribal groups in the south west region alone, each with varying dialects.

We know that many plants such as Eucalypts - Wandoo, Jarrah and Marri, derived their common name from the translations of the south west Aboriginal group the Nyoongar people.

Listed below are just a few of those names recorded by early settlers and historians.

References and suggested reading

Beard, J.S.   *Plant life of Western Australia..* 1990 Kangaroo Press. NSW
Bennett, E.M.   *Common and Aboriginal names of Western Australian Plant Species.* 1995
                                     Wildflower Society of Western Australia. Perth. WA
Bennet, E.& Dundas, P.   *The Bushland Plants of Kings Park Western Australia.*
Brooker, M.H. & Kleinig, D.A.   *Field Guide to Eucalypts. South-western and southern Australia.* 1990. Inkata Press.
Clarke, I.& Lee, H.   *Name That Flower.* 1997. Melbourne University Press. Carlton. VIC.
Corrick M. & Fuhrer,B.   *Wildflowers of Southern Western Australia.* 1996. The Five Mile Press Pty Ltd.
Craig, G.   *Native Plants of the Ravensthorpe Region.* 1995. Ravensthorpe Wildflower Show Inc.
Erickson, George, Marchant & Morcombe   *Flowers and Plants of Western Australia.* 1986. Reed Books Pty Ltd.
Flannery, T.   *The Future Eaters.* Reed Books.
George, A.   *The Banksia Book.* 1996. Kangaroo Press. Sydney. N.S.W.
Green, J.W.   *Census of the Vascular Plants of Western Australia.* WA Herbarium.
Hoffman, N. & Brown, A.   *Orchids of South West Australia.* 1992.University of Western Australia. Perth. W.A.
Hopper, S.D.   *Kangaroo Paws & Catspaws.* 1993. Department of Conservation and Land Management.
Hopper, Chappill, Harvey   *Gondwanan Heritage, past Present and Future of Western Australian Biota.* Surrey
and George, editors.   Beatty and Sons.
Hopper,van Leeuwin, Brown and Patrick. *WesternAustralia's Endangered Flora.* Department of Conservation and Land
                                     Management.
Marchant, Wheeler, Rye,   *Flowers of the Perth Region Vol. I & 2.* 1987. Western Herbarium. Perth. WA
Bennett, Lander & McFarlane.
Olde, P. & Marriott,N.   *The Grevillea Book. Vol. 1.2.& 3.* 1995. Kangaroo Press. Sydney. NSW
Sainsbury, R.M.   A Field Guide to Dryandra. 1985. A.Field Guide to Smokebushes & Honeysuckles. 1991.
                                     A Field Guide to Isopogons & Petrophiles.1987. University of Western Australia. Perth.WA
Saunders Craig and Mattiske.   *The Role of Networks in Nature Conservation.* Surrey, Beatty and Sons.
Sharr, F.A.   *Western Australian Plant Names and their meaning.* UWA Press.
White, Mary E.   *The Greening of Gondwana.* Reed Books.

Science papers
Chapman, A. and Newbey, K.R.   *A biological survey of the Fitzgerald area, Western Australia.* CALM Science Supplement
                                     Three 1979.
Hopper, S.D.   *Biiogeographical Aspects of Speciation in the Southwest Australian Flora.* Annual
                                     Reviews Incorporated.1993.
Pate, J. S. and Hopper, S.D.   *Rare and Common Plants in Ecosystems, with Special reference to the South-west
                                     Australian Flora*

Guide to the Wildflowers of South Western Australia
Published by Simon Nevill Publications. Unit 7 - 342 South Terrace South Fremantle Perth WA 6162 Australia.
Telephone 089 336 3882.   Fax. 089 336 3882.   Email - falcon @ highway 1. com . au
Nevill.Simon.J. Copyright 1998.
Guide to the Wildflowers of South Western Australia.
ISBN. 0 9585367 08.

Photography, Original book concept, map drawings, book design and computer layout - **Simon Nevill.**
Text: Nathan McQuoid.
Section -   'The Wildflowers of the South West'.   'The Botanical Zones of the South West Botanical Province'
Text: Simon Nevill.
Section -   'Where and When to find Wildflowers in the South West'   'Plant names & Plant structure'
        'Landcare'   'Chapter 3 Plant names. locations. flowering times.'
Text: David Knowles.
Section -   'Pollination & Honeybees'

Text: Peter Smith.
Section -   'Growing Native Plants'
Computer graphics & final book layout - **Rob Leming.**Perth. **Leanne Quince.**of 'Graphics Above' for computer typography.
Reprographics - **Colourbox Digital.** Perth.   Printing by **Kyodo Printing Co. Pty. Ltd.** Singapore.
Published by Simon Nevill Publications. Perth Western Australia.

All rights reserved. No part of the publication may be reproduced, stored in a retrieval system or transmitted in any form or by any means, mechanical, by photocopying, electronic, recording or otherwise without the consent of the publisher.

Information update. During 1997 between tours, over 45,000 kil. were travelled in the South West, to photograph the 900 + species illustrated in this book. To achieve this, over 5000 slides were taken. With such a large quantity of species illustrated in one volume, it is possible to have an incorrect identification of a given species. The authors welcome feedback in writing from readers that may correct information given or establish additional information regarding flowering times or extend the distribution of plants from those botanical zones stated. This will assist in keeping future editions up to date.

Photo by Marcelo Palacios

Simon Nevill

For over 10 years now Simon has been running Falcon Tours, a wildlife tour company with his main expertise being in the field of ornithology. Having seen the vast majority of Australian birds, he needed a new challenge. He never realised it would be at his own feet here in Western Australia. It of course is the wonderful world of wildflowers. He says "I'm not a botanist but I wanted to share my new found passion with others, as I felt there was a need for such a book."

Nathan McQuoid

Nathan McQuoid was introduced to nature by his grandfather Bert Legg in the 1960's. The influence was profound in developing his passion for Western Australian nature and in shaping his future. Since then he has enjoyed a career involved in the conservation of nature across some of the most significant landscapes in South Western Australia. His interests are in the improved understanding of broader landscape biogeography and ecology, and in particular the natural distribution of the eucalypts and their use in ecological restoration.He was also fortunate to have the opportunity to have travelled with and be influenced by George Gardner and the late Ken Newbey.

Acknowledgements

One of the joys of producing a book of this nature, is that it becomes a team effort involving like minded people though special thanks must go to Nathan McQuoid for his contribution to the text and for assisting with field information and book concepts but most of all, for sharing his enthusiasm and love of the plant world.
To Peter Smith proprietor of Seed West who contributed to the section on growing native plants, assisting with plant identification and simply being a good friend.   To David Knowles for contributing the section on pollination and sharing his knowlege of the world of entomology.   To Dr Stephen Hopper for reviewing the final manuscript.   To Dr. Eleanor Bennett for assisting with plant identification and checking the manuscript. To the WA Herbarium for the kind use of their reference herbarium under Dr Neville Marchant with specific help on Acacias from Dr Bruce Maslin.   Also to Dr Jenny Chappell of the UWA, Bob and Barbara Backhouse for assisting with plant identification.   To Rob Leming for his contribution to the computer graphics and lay up of this book and for burning the midnight oil. To Leanne Quince for her assistance with typography and computer help.   To Kim Burket for her hard work in filing my plant records.   Dr. Christine Newell of the UK for being a wonderful travelling companion. To my friends Brenda Newbey , Ian Edwards and Graham Jones for assisting with text. To the staff at Colour Box Digital - Perth, who put up with the needs of a perfectionist.   To Cliff and Dawn Frith of Queensland, authors of many books, who as true friends gave unselfishly of their publishing knowledge. Finally to my parents Nim and Mary who both celebrate their 89th birthdays this year, this book is dedicated to you. To anyone I have overlooked, my sincere apologies.

Simon J. Nevill. April 1998.